Also by Richard Restak, M.D.

PREMEDITATED MAN:
BIOETHICS AND THE CONTROL OF FUTURE HUMAN LIFE

THE BRAIN: THE LAST FRONTIER

THE SELF-SEEKERS

THE BRAIN

THE INFANT MIND

THE MIND

THE BRAIN HAS A MIND OF ITS OWN

RECEPTORS

THE MODULAR BRAIN

BRAIN

How New Discoveries in Neuroscience
Are Answering Age-Old Questions
about Memory, Free Will, Consciousness,
and Personal Identity

Richard M. Restak, M.D.

A TOUCHSTONE BOOK
Published by Simon & Schuster
NEW YORK   LONDON   TORONTO   SYDNEY   TOKYO   SINGAPORE

TOUCHSTONE
Rockefeller Center
1230 Avenue of the Americas
New York, NY 10020

First Touchstone Edition 1995

TOUCHSTONE and colophon are registered trademarks
of Simon & Schuster Inc.

Manufactured in the United States of America

10 9 8 7 6 5 4 3 2 1

Library of Congress Cataloging-in-Publication Data
Restak, Richard M., date
    The modular brain : how new discoveries in neuroscience are answering age-old questions
about memory, free will, consciousness, and personal identity / Richard M. Restak.
        p.   cm.
    Includes bibliographical references and index.
    1. Neuropsychology.   2. Brain.   I. Title.
QP360.R455   1994
153—dc20   93-42298
                                                                        CIP

ISBN 0-684-19544-5
     0-684-80126-4 (Pbk)

Lecture by Roger W. Sperry entitled "Consciousness, Personal Identity, and the Divided
Brain" reprinted by permission of the author.

To Ann Buchwald
for her help and support through the years

CONTENTS

ACKNOWLEDGMENTS

Thanks are due to the following who generously provided interviews, guidelines, suggestions, and richly rewarding conversations:

Alfonso Caramazza, Ph.D., cognitive scientist at Johns Hopkins University; Joseph E. LeDoux of The Center for Neural Science, New York University; Barry Gordon, M.D., Ph.D., neurologist and cognitive scientist at Johns Hopkins University; Jordan Grafman, Ph.D., Chief, Cognitive Neuroscience Section, National Institutes of Health; John Hart, Jr., M.D., cognitive scientist at Johns Hopkins University; Kenneth M. Heilman, M.D., the James E. Rooks, Jr., Professor of Neurology, University of Florida College of Medicine, Gainesville, Florida; Mortimer Mishkin, Ph.D., Laboratory of Neuropsychology, National Institute of Mental Health; Vernon B. Mountcastle, M.D., University Professor of Neuroscience at Johns Hopkins University; Guy McKhann, M.D., director of the Zanvyl Krieger Mind/Brain Institute at Johns Hopkins University; Paul McHugh, M.D., Henry Phipps Professor of Psychiatry and Behavioral Sciences at the Johns Hopkins School of Medicine; and Rachel Wilder, formerly of the Office of Public Affairs, Johns Hopkins Medical Institutions.

INTRODUCTION

At home I have a diagram taken from a children's encyclopedia from the 1930s. It's labeled "The Central Control Station of Your Body," and shows a cutaway view of a human head containing tiny rooms, occupied by one or more men dressed up in business suits fashionable at the time.

In one room, a man is sitting at a desk talking into one of those two-piece telephones, the earphone held in his hand and the mouthpiece sitting in front of him on the desk. Written above this scene is the phrase, "Manager of Speech." Toward the back of the brain, a small box houses a man scanning pictures on a desk. He is the "Receiver of the Camera Pictures," forwarded to him over "air tubes" from a room farthest to the front containing four men dressed in sailor uniforms and pulling on rigs and pulleys controlling the eye (they are identified as the "Camera Operators"). At the center of the cutaway head is a room larger than any of the others and occupied by three men sitting at a conference table (in line with the prevailing practices and prejudices of the time, only white men are depicted in charge of the "control stations"). This caption reads, "Brain Headquarters in the Cerebrum."

Altogether nine tiny boxes filled with men carrying out various tasks illustrate the workings of the brain. The caption for the diagram as a whole reads: "Imagine your brain as the executive branch of a big business. It is divided as you can see here into many departments. Seated at the big desk at the headquarters office is the General Manager—your Conscious Self—with telephone lines running to all departments."

Today it is easy to ridicule such a simplistic view of how the brain works. We don't experience ourselves as a "General Manager" issuing orders to members of a bureaucratic organization. Greeting a friend, for instance, doesn't seem to have anything in common with the process described in this fanciful diagram: "Instantly you begin issuing orders: 'Tell the Speech Manager to say "Hello Johnny!" Tell the Leg Superintendent to stop walking at once! Tell the Arm Superintendent to stick out my hand right away and take Jones's hand! Tell the Face and Lips Superintendent to give this man a good big smile!'"

Rather, most of the "orders" issued and received within the body must be unconscious ones, since we are not aware of issuing them. Indeed, most of the operations of our brain take place outside of our conscious awareness, and on the whole that isn't such a bad thing. None of us would wish consciously to have to initiate every action, no matter how petty, or to be aware of even the most trivial of sensations.

But at the highest levels of our mind's operations, during those less pressured, more "philosophical" moments when we delve inward and explore the terrain of our own minds, we tend to encounter and experience a "self," a conscious controller who wills, remembers, decides to act, experiences emotions, feels pain, and simultaneously rejoices and fears for the future. And at such times we tend to subscribe to something similar to the General Manager in the diagram. But when we try to pin down this "self," this General Manager of the psyche, our experience is like trying to ensnare a fish that has caught our eye by reaching into a pellucid lake with our bare hands. We *experience* consciousness and awareness in ourselves, but have no way of conveying these experiences to others. It's not that we doubt—except in moments of intoxication, fever, or insanity—the reality and existence of our inner self, our "mind." It's just that, try as we might, we cannot pin it down in regard to its exact nature or location.

For centuries, the mind was treated as a mysterious, ethereal entity that could not be further inquired about and the brain, the originator of all thought, was either ignored altogether or treated as an inconvenient irrelevance. Such an attitude was at least partly based, I am convinced, on a kind of "turf battle." Traditionally, "why" questions (Why am I here? Why is the world the way it is?) have been asked by philosophers, and until very recently, few philosophers expressed any interest in learning about the brain. As a result, many of the most interesting questions were approached in a false and artificial way that turned off many people, myself included, who find philosophical questions fascinating and challenging.

In my elementary philosophy class, for instance, we spent hours debating whether a tree falling in a forest empty of people made any noise. The solution to this conundrum comes not from the use of syllogisms or other forms of reasoning, but from knowledge about the human ear's connections to the brain along the auditory pathway from the tympanic membrane to the medial geniculus in the midbrain, to the primary auditory area in the temporal lobe, and finally to the auditory association area, where the sound is recognized and identified. If the sound waves created by the falling tree fail to strike a tympanic membrane, or for some reason that message is not relayed onward to the brain, then no sound exists; since sound, by definition, requires an ear and a brain. Absent these, we have only waves of a certain frequency.

Another much debated question has to do with the reality of space and time independent of the mind. The philosopher Immanuel Kant, who knew nothing about the brain, arrived intuitively at the conclusion brain scientists would reach two hundred years later. In *The Critique of Pure Reason*, published in 1781, he wrote that space and time corresponded not to "objective" realities existing independently of the mind (we would now say the brain) but are themselves categories *created* by the mind. He claimed we can have no knowledge of things as they are in themselves, existing in a strictly independent physical sense. This is because our senses construct our world for us and therefore do not present to the mind external independent realities but only perceptions. In essence, our "objective" world is a highly subjective one.

When I first encountered Kant I found his ideas appealing, even though it would be another decade before I confirmed that Kant was right when I encountered neurological and psychiatric patients who suffered from disabilities affecting their appreciation of space and time. Now, after twenty years studying the brain and writing eight other books about it, I think the time is right to explore some traditional philosophical questions in the light of what we have learned about the brain over the past several decades.

Consciousness, thought, memory, will, emotion—none of these has any independent outside reality other than in the context of the human brain. All are based on the brain's organization. This concept takes some getting used to at first. Few of us think of ourselves in terms of our brains. If we think about the brain at all, it is usually in a way similar to the way we think about the rest of our body. For instance, we may take up dietary approaches aimed at increasing our alertness or mental efficiency. To this extent the brain is just like every other organ composing the great "machine," the human body. But few of us equate ourselves with our brain. Yet a few moments spent with a person who has just suffered a stroke illustrates in a stark and vivid manner one of nature's aphorisms: Change the brain, change the person. As a result of brain injury, the person is not the same person as before. Insofar as we can enter into the stroke victim's mental life—a task we will attempt at various points throughout this book—we encounter clear indications that perceptions, thoughts, and emotions are drastically different than before the brain injury. What's more, these differences are often unexpected, even bizarre. In the pages that follow, for instance, we will encounter a man who loses the ability to name animals but has no difficulty in naming inanimate objects; another who can recognize tools but not musical instruments. Such surprising impairments suggest that the brain is organized differently than was believed by most experts in the past. Who could have ever imagined, for instance, that the concept of size could exist independently within the brain, so that a person could lose the sense of whether an elephant is bigger than a mouse but remain normal in every other way? None of these strange and tragic impairments could have been predicted. Nor could we learn what they teach us about the reality formed for us by our brain by simply sitting around and thinking about them or discussing them with others. Logical analysis, reason, or the other methods favored by philosophers are little help here. Only patients with these impairments, many of them severely disabling and all of them tragic, can teach us how our brain constructs our individual and collective "reality." Nature is the teacher, the patient the "experiment," and the human brain the classroom.

The Modular Brain describes recent discoveries about brain organization and what that means for our ideas about the nature of reality. Thus the book is in the tradition of exploring the "mind–body problem." But I prefer to think of it as finding and identifying the General Manager depicted in our 1930s diagram, if he exists, or permanently firing him if he doesn't. Let's start with a quick overview of how we will proceed.

History provides three revolutionary insights into the relationship of the brain and the mind.

The first insight dates to the Greeks, specifically the physician and philosopher Hippocrates, who wrote:

> Not only our pleasure, our joy and our laughter but also our sorrow, pain, grief and tears arise from the brain, and the brain alone. With it we

think and understand, see and hear, and we discriminate between the
ugly and the beautiful, between what is pleasant and what is unpleasant
and between good and evil.

With his recognition of the brain as the seat of the mind, Hippocrates
overthrew thousands of years of speculation favoring other bodily
organs. (The Egyptians selected the liver; Aristotle held out for the
heart, considering the brain little more than a cooling system for the
blood.)

Two thousand years later, two early nineteenth-century European
physicians, Paul Broca and Carl Wernicke, conducted or supervised
autopsies of patients who during their lifetimes had suffered from aphasia, an impairment of their capacity to speak or understand language.
Broca and Wernicke discovered areas of brain destruction in the left
hemisphere of these patients. From this came the second revolutionary
insight. Broca and Wernicke concluded that the left hemisphere mediated the production and understanding of spoken language. But their
findings had a wider application than just language. In addition, they
provided support for the view that the brain is not a homogeneous
structure in which one part is equivalent to every other but, instead, is
made up of special centers located at various locations in the two hemispheres.

Following on the discovery of the language centers came the discovery of the visual areas toward the back, hearing along the sides, and
movement and sensation toward the front. Over the ensuing 120 years,
neurologists and others have established additional correlations between
brain structure and function.

The third revolution in our understanding of the brain extends back
not more than a few decades, with much of the important work no
more than a few years old. It evolved from the view that the brain is
organized according to a hierarchical structure wherein the cerebral
hemispheres, the most recent and highly evolved areas of the brain,
control those more ancient areas inherited pretty much unchanged from
our evolutionary ancestors. While this view is still useful in its way, it
ignores recent evidence that even such a seemingly straightforward
function as vision is not at all a unitary process but involves separate
parallel processes that, working together, result in our visual experience.

The Modular Brain presents this new way of thinking about our
mind and its relationship to our brain. It is organized into three sections.

First, we will examine what has been thought over the centuries and until very recently about the relationship of mind and brain. When we compare what we know about the brain now with the thoughts held by our predecessors, it's easy to become smug and assume we will soon learn everything there is to know about it. But that smugness soon disappears when we discover that although we have learned more about the brain in the last one hundred years than in all previously recorded history, we remain woefully ignorant of the bases for most of the operations of the brain. Moreover, as with physics after the introduction of quantum mechanics, neuroscience has still to come up with a unified theory that incorporates new discoveries about perception, thought, memory, and other brain processes. Sixty years ago, the Nobel prizewinning neuroanatomist Santiago Ramon y Cajal wrote: "In truth it is not possible, in our present state of knowledge, to formulate a definitive theory of the functional plan of the brain." If we emphasize the word *definitive*, then Cajal's sentence is as true today as the day he wrote it. But we have learned a lot in the intervening sixty years and can now offer some intriguing ideas about how our brain works.

In the second part, we will explore research leading up to this new truly revolutionary view of the brain's operation, modular theory. Briefly, this theory holds that our experience is not a matter of combining at one master site within the brain all of separate components into one central perception. As strange as it may sound, there is no master site, no center of convergence. For centuries we thought there had to be. Descartes and other philosophers referred to the brain's central coordinating area as a "homunculus," a personified character much like the General Manager, a kind of ghostlike figure later satirized by philosopher Gilbert Ryle as "the ghost in the machine." Ryle's point in referring to a ghost was, of course, that there is no evidence for such a central autonomous overseer within the brain. Besides, if such a creature existed, who would be monitoring him or her (or it)?

Wherever one looks in the brain—particularly the cerebral cortex, the overarching expansion of the brain that in size, complexity, and cell number differentiates us from all other living things—all brain cells and collections of brain cells communicate with other cells. This means that no "pontifical" cell or area holds sway over all others, nor do all areas of the brain "report" to an overall supervisory center. Thus, to betray my position at this early point in the game, the General Manager is a fictional character. And like all fictional characters he was created in order to fill a need within a story, in this case the early efforts to come up with a theory of how the brain works. But with a change in the plot, the characters too must yield place to different characters.

A theory based on multiple connections all operating simultaneously and in parallel has profound implications. "It implies that there is no cortical terminus, no final destination where the soul or consciousness, for example, may reside," according to Semir Zeki, a neuroscientist who has played a leading role in the research establishing the modular theory of brain organization.

Finally, we will explore what a modular organization of the brain means for our understanding of the human mind. If no "final terminus" or center exists within the brain, then who is this "me" encountered in moments of introspection? Most of us certainly don't inwardly experience ourselves as modular, but as singular and reasonably unified. Yet, as we shall see, important select aspects of ourselves may be altered or disappear altogether on the basis of damage to our brain (in other instances, of brain maldevelopment, they may never have evolved in the first place). If I am the result of the operation of a vast network of nerve cell interconnections, in which one cell may be influencing the response of any of the other 40 to 50 billion other cells, then what attitude should I take to my inner perception that I remain basically the same person throughout my life? What of free will?

Philosophers have talked and debated for centuries about the unity of the personality, the nature of emotion, autonomy, and creativity. If you want to see how much they have accomplished and how little they agree about any of these terms, open a philosophy textbook. Not only is there no agreement, but many of the authors of these works seem to take a perverse pride in compiling long lists of what each separate philosopher thought and taught about even such seemingly elementary mental processes as perception. One reason for this smorgasbord approach to the mind is that—contrary to what most philosophers have written—perceiving and knowing are not separate processes, with one following the other like a bucket of water passed from hand to hand along a chain of firefighters, but *one process*. Vision, for instance, is not at all like a camera where the eye focuses light on the retina where a "picture" is taken and conveyed along the visual pathways to the brain for interpretation. Rather, the brain actively constructs what we "see," and we are at once camera, film, photographer, and picture. Just as no two pictures are ever exactly the same because of differences in one or other of these elements, so too our brain, thanks to its modular organization, creates for each of us a unique "world picture."

In the pages that follow, we will explore the implications of modular theory for our ideas about memory, consciousness, free will, and personal identity. Modular theory helps us to set to rest many old puzzles and conundrums, but in the process creates some new and unique ones. To mention just one, artistic creativity involves some brain areas more than others; these areas also seem to be more developed in people proficient in the arts. Does this mean that artistic creativity is primarily an inherited trait, and that those with the "wrong" genes cannot hope to be artistically creative? Everyday observation seems to contradict that claim: Many artists have come from families with unremarkable artistic abilities. Yet artistic creativity of some artists at the highest levels seems to involve mental abilities (perfect pitch, writing and "listening" to musical compositions in one's head) that involve more than simple refinements on commonly encountered musical talents. Ravel was such an artist. His diaries describe his creative process and how that process broke down as a result of brain damage. Recent experiments employing PET scan studies of musicians have shed additional light on Ravel's creativity and modular brain mechanisms.

It is my hope that *The Modular Brain* will inspire general readers to consider the implications of this new theory.

THE MODULAR BRAIN

CHAPTER ONE

"On the Organ of the Soul"

n its natural state the brain defies our attempts to come to grips with it. Moments after death and removal from the skull, it begins to collapse and deteriorate into a viscous jelly. Only if suspended in a chemical—the ancients favored wine—can it retain its shape. But even then the most careful observation takes in no more than its most obvious anatomical boundaries. Additional knowledge of the brain's organization awaited the development of more specific chemical methods capable of "fixing" the brain and revealing its separate components.

One neuroscientist once compared the brain to an archeological dig: the more ancient, "primitive" areas located deep beneath the surface; the more recently evolved structures (the cerebral hemispheres) located toward the surface. While such a conceptual framework is useful, it is also limiting and betrays a certain chauvinism. From our particular vantage point, a brain designed for language along with the development of tools and technology is superior and "higher" than one designed for other functions. But if we can step back from ourselves, try to see things in a less self-referential way, we realize that "higher" and "lower" are merely terms we apply to distinguish those creatures closest to ourselves ("higher") from those quite unlike us (the "lower" forms). The brain of a dolphin is specialized so as to favor enhancement of brain centers useful for survival in the ocean. Thus the dolphin brain is "higher" and more evolved than our own in an aquatic environment, but "lower" and not designed at all for survival on a large metropolitan freeway. Rather than thinking of creatures as higher or lower, it's more useful to think of them as encountering different "realities" created for them by the organization of their brains. With this important caveat in mind, let's take an overview of the general organization of the human brain.

Since our concepts of the brain are continually being revised, the following is not immutable and set in stone but subject to revision depending on future discoveries. For instance, the cerebellum, the cluster of nerve cells and nuclei at the very back of the brain, was considered until recently important only for the carrying out of smooth coordinated movements. It was thought to function as a kind of error-measuring device that compares the motions and movements actually performed with the intended action. While that remains true, recent research reveals that the cerebellum is also involved in something more properly philosophical: the carrying out of willed acts and the control of hostile impulses. Thus a "lower" portion of the brain, well below the cerebral hemispheres that distinguish our species, has come to be identified after additional discoveries with "higher" functions. We will encounter other reassignments and redefinitions of function as we progress but, since we must start somewhere and I generally subscribe to the view that traditions are to be respected until disproved or rendered obsolete, let's start with the generally accepted ideas about the brain and its organization.

The brain is actually an enlargement and opening outward of the spinal cord (see diagram 1). Like the spinal cord, it consists of grey matter made up of the cell bodies and dendrites of the neurons (brain cells) and white matter made up of tracts connecting parts of the brain with each other. The first structure encountered as one ascends upward from the spinal cord is the slightly enlarged medulla oblongata. It contains clusters of neurons (nuclei) which control such basic biological processes as blood pressure, heart rate, and breathing. Next comes the other component of the hindbrain, the pons (from the Latin word for "bridge"). It serves as a conduit for fibers from one side of the hindbrain to the cerebellum on the opposite side. Together the pons and the medulla oblongata contain the nuclei of origin for the fifth through the twelfth cranial nerves which are responsible among other things for sensation to the face and facial movement, outward movements of the two eyes, taste, hearing, swallowing, and tongue movement. Ascending still further one encounters the midbrain, the point of origin for the third and fourth cranial nerves which move the eyes (both third and fourth) and control the size of the pupil (the third). Still further upward is the forebrain or cerebrum made up of the two cerebral hemispheres.

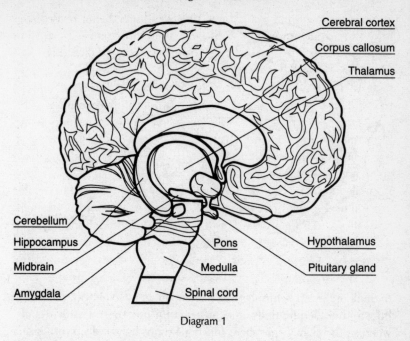

Diagram 1

Each cerebral hemisphere is divided into lobes: the frontal, temporal, parietal, and occipital (see diagram 2). These divisions are artificial—the brain is organized and functions as a whole—and based on arbitrary but generally agreed upon anatomical landmarks. The divisions are also of functional importance. The occipital lobe is concerned with vision, the parietal lobe with sensation from all of the senses except smell, the temporal lobe with hearing and memory, the frontal lobes with social behavior and such personal characteristics as curiosity, foresight, and the capacity to foresee the consequences of one's actions.

Each cerebral hemisphere directs the movements of the arm and the leg on the opposite side of the body. Coordination of the hemispheres is brought about by fibers passing across the corpus callosum, an eight-million-fiber bridge of nerve fibers. Despite the superficial appearance of similarity, the two hemispheres differ both in form and activity. A structure in the temporal lobe, the planum temporale, is usually larger on the left, and this structural difference is evident as early as foetal life. The functional differences, the so-called "right-left" brain differences, will be discussed in more detail in Chapter Ten. For now it is sufficient

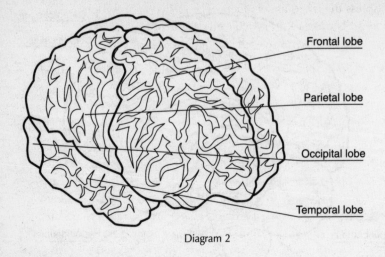

Frontal lobe

Parietal lobe

Occipital lobe

Temporal lobe

Diagram 2

to think of the left hemisphere as working in an analytic way, processing information sequentially and abstracting out the relevant details, whereas the right hemisphere functions more holistically, synthesizing many myriad details into a unity. Taking verbal directions about how to get from one place to another involves the left hemisphere, whereas looking quickly at a map and taking in the route as a whole activates the right hemisphere.

The outer layer of the two hemispheres is the cerebral cortex or "bark" (from the Greek), which appeared late in evolution and is especially developed in humans. Since this is the part of the brain that will be our main focus, I will present it in more detail.

Only about 2–5 millimeters in thickness, the cortex consists of some twelve billion neurons and their cell bodies and branches along with an unknown number of glial cells (from the Greek for "glue") which are supportive to the neurons. Estimates of the number of neurons in the human brain vary widely according to whether or not the neurons of the cerebellum are included (they are smaller and much more difficult to count). But estimates of the total cell number provides only limited information about the brain, since at any given moment thousands if not millions of neurons are altering their connections with each other. Above all, the brain is a dynamic organ, so dynamic in fact that as I write that word "dynamic" I am aware that the word doesn't adequately express the true state of affairs. The pioneering neurophysiologist Sir Charles Sherrington, in a frequently quoted passage, came as close to describing the dynamism of the human brain as anyone:

> It is as if the Milky Way entered upon some cosmic dance. Swiftly the brain becomes an enchanted loom where millions of flashing shuttles weave a dissolving pattern, always a meaningful pattern though never an abiding one; a shifting harmony of subpatterns.

Beneath the cortex are islands of gray matter, the subcortical nuclei, which relay impulses to and from the overlying cerebral cortex. These areas, with names such as the globus pallidus, the putamen, and the caudate nucleus, aid in controlling and modulating movement. Parkinson's disease, which is marked by slow movements and tremor, results from a dysfunction of nerve fibers terminating on a group of these subnuclei collectively called the corpus striatum (another term taken from Latin, meaning "striped body"). The largest and most important subnuclear way station—this one for impulses ascending *to* the cortex—is the thalamus. Taken from the Greek word for "inner room" traditionally applied to a bridal chamber, the thalamus sits just outside the main entrance to the cortex. It is subdivided into a dozen or so regions, each of which relays impulses to different parts of the cerebral cortex.

Immediately beneath the thalamus is the hypothalamus. The most powerful portion of the brain, occupying less than one percent of its total volume, the hypothalamus organizes metabolism, hormone circulation, heat production and dissipation and body temperature control ("heat stroke" results from a breakdown in this function), internal states such as alertness or sleep, and aggression or timidity (we will have more to say about this last function later). It also is involved in sexuality and the operation of the autonomic nervous system.

At the back of the head is the cerebellum, mentioned above. Highly developed in certain fish, sharks and birds, the structure makes possible delicate and skilled movements. Examples are an osprey soaring from the sky to snare a fish at just the right moment from beneath the surface of the water or, in our own species, the delicate pirouette of the ballerina or the split-second timing of the knockout punch of the champion boxer.

Finally, at centers in the brain stem and spinal cord are the elements of the autonomic nervous system. Its name dates from the notion that its processes cannot be brought under voluntary control. But modern biofeedback techniques disprove this; thousands of people with hypertension and migraine headache have learned to control the physiological mechanisms responsible for these disorders. A truer statement would be that the actions of the autonomic nervous system rarely enter consciousness and when they do, the end result is more often harmful than helpful to one's general health. Most of us are mercifully unconscious, for instance, of the so-called piloerector response whereby the hair on our arms and hands stands on end. We experience the fright associated with this response and we can literally feel the hair standing upright, but we don't consciously *experience* their arising.

At the lower microscopic level of description, the brain is made up of specialized cells, neurons, consisting of a cell body or soma that gives rise to two types of processes—the *axon* or nerve fiber which transmits the impulse from one cell to the *dendrite* of another cell which receives it. With somewhere in the neighborhood of 10 billion nerve cells, some of which make contact with 10,000 others, the number of possible nerve cell interactions exceeds the estimated number of particles of matter in the universe! Despite the neuron's importance, our knowledge of it dates back no more than 150 years (Charles Darwin knew nothing of neurons, and in fact spoke little of the brain even though he considered it the evolved organ of the mind). Only with the development of special chemical stains (at first silver or gold in the 1880s, followed later by more prosaic substances like organic dyes) could the neuron be seen under the microscope. Some nerve fibers are encased in a sheath of fat and protein called myelin which speeds up transmission. These fibers with their glistening appearance form the white matter of the brain which on staining is easily distinguishable from the gray matter comprising the nerve cell bodies.

Neurons are electrically excitable cells resulting from separation of ions (charged particles) across the membrane that encloses the neuron. This separation of ions is maintained by differences in permeability of the membrane to different ions. It is very permeable to potassium ions, less permeable to chloride ions, and much less permeable to sodium ions. These differences in permeability across the membrane result in different concentrations of ions on the inside and the outside. This uneven concentration produces a difference in electrical potential, the outside electropositive, the inside electronegative. Whenever the membrane is stimulated, a small region at the site of stimulation allows sodium ions to rush in. This creates the nerve impulse, which travels along the length of an axon or dendrite in the form of waves of altered membrane potential. Known as the "action potential," the nerve impulse is self-propagating and flows without decrement in one direction only.

When the nerve impulse reaches the end of the axon, it communicates with the dendrite of another neuron at the special junction, the synapse (see diagram 3). There the message is transformed from an electrical to a chemical one. Depolarization of the first cell, the presynaptic axon terminal, brings about the release of a chemical substance, the neurotransmitter, which crosses the synapse and attaches to a specialized receptor for that chemical on the dendrite of the second nerve cell (the postsynaptic neuron). For our purposes it is sufficient to note that different neurons use one or more of the fifty to one hundred different neurotransmitters. Some are excitatory and enhance the likelihood that a neuron will stimulate another neuron to "fire," while others are inhibitory and lessen the likelihood of that happening. The situation is like what happens at one of those New England town meetings where the final decision to take or not take some action depends on the number of voices or votes on each side of the question. If there are more inhibitory influences on the neuron, it will not become depolarized enough to "fire"; if excitatory influences predominate, the necessary change in membrane permeability will occur and an action potential will be generated. The brain as a whole is best understood as made up of the sum total of excitatory and inhibitory neuronal interactions at any given moment. This balance is always unique throughout our lifetime, one of the reasons for the dynamic nature of the brain. Not only are the brains of different people always unique—even those of identical twins—but an individual brain never exactly repeats itself, never expresses exactly the same balance of excitation and inhibition.

At this point you know more about the brain than the wisest and most informed persons over the previous two millennia. Based on this, it is easy to conclude that we have nothing to learn from our forebears. But this is a mistaken notion since their efforts to understand the brain resulted in some ideas and concepts that continue to shape our own attitudes today both for good and for bad. I have already mentioned the "higher-lower" distinction, based on the notion that thinking and willing are characteristic of no other creatures than ourselves and therefore

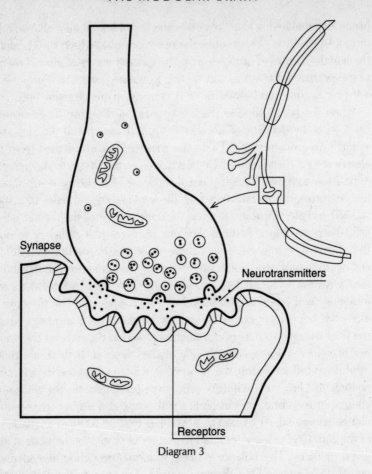

Synapse

Neurotransmitters

Receptors

Diagram 3

must emanate from "higher" brain areas (in both the physical and metaphoric sense), while other activities that we share with other animals can be accounted for by "lower" centers. But the influence of previous thinkers—many of them philosophers—is even more fundamental. Consider, for instance, the quest to determine which parts of the brain house the human spirit or soul.

I mentioned a few pages back that the softness of the unfixed brain precluded any examination of its structure. For this reason, the earliest observers who had no method of fixing it concentrated on the ventricles, those well demarcated cavities within each cerebral hemisphere and the upper brainstem. The contents of the ventricles, air according to some, fluid according to others, seemed a plausible vehicle for transporting the "animal spirits," a term that endured from several centuries B.C. up to the eighteenth century as a designation for the unit of nervous transmission. From this emphasis on ventricles, or "cells" as they were oftentimes referred to, arose the notions of receptacle, reservoir, and circulation.

According to one expression of the theory, a human being is made up of a "vital principle" or spirit with its place in the ventricles (hollows) of the brain, from which it controls the senses. From the ventricles emerge the nerves, which serve as pipes to conduct the "sense-producing spirit" to appropriate destinations. The brain substance itself, the gray matter or cerebral cortex, was relegated to the status of a gland which secreted the fluid contained within the ventricles. The white matter with its many "canals" visible with magnification served only as a transporter of that fluid to the nerves throughout the body which, in turn, conveyed it to the muscles. Behind this model lay an implicit analogy between a temple and a brain.

Ancient temples consisted of the vestibulum, consistorium, and apotheka. In the vestibulum, petitioners submitted legal declarations; in the consistorium, authorities scrutinized these statements; and in the apotheka, subsequent decisions and sentences were pronounced. Contemporary authorities attributed similar functions to the cerebral cavities. Information gathering in the anterior part of the brain contrasted with executive functions carried out posteriorly. Diagrams of the head and brain dating from the Middle Ages identify anterior, middle, and posterior ventricles, held to be responsible for, respectively, imagination, reason, and memory—the three generally agreed-upon components of the intellect.

Accompanying these early anatomical correlations came references to moisture, temperature, and consistency. All substances, according to the philosopher Anaxagoras, consisted of pairs of opposite qualities: wet/dry, hot/cold, light/dark. Physicians applied these theories to health and disease. Alcmaeon of Croton decked out this primitive theory of the brain with a political metaphor: "Health results from equal rights of the qualities, wet, dry, cold, hot, bitter, sweet, etc. A dictatorship among the qualities produces disease."

Vision and fantasy were thought to be housed within the dry and warm first ventricle, logic and discerning reason in the warm and moist middle ventricle, and memory in the dry and cool posterior ventricle. But whatever the purported functions of the ventricles—and there were many—their chief function was to hold and distribute what the sixteenth-century French mathematician, philosopher, and physiologist René Descartes referred to as the "subtle, volatile, invisible, and immaterial animal spirits." Shakespeare provided poetic form for this doctrine of the "animal spirits." "This is a gift I have," says Holofernes in *Love's Labour's Lost* (act 4, scene 2), "full of forms, figures, shapes, objects, ideas, apprehensions, motions, resolutions: these are begot in the ventricle of memory, nourish'd in the womb of pia mater, and deliver'd upon the mellowing of occasion."

The belief in "animal spirits" coincided with the goal of locating within the body the seat of the immortal soul. For this reason, the earliest efforts at understanding the relationship of what we would now call the mind to the brain involved theologians rather than physicians or scientists. Indeed the clergy dictated anatomical concepts during the medieval period. An illustration of the brain dating from 1496 and now in the Wellcome Historical Museum in London includes two versions of brain organization from which the interested observer can choose. At the upper left is shown a four-celled brain representing the teachings of the ancient physicians Galen and Avicenna, and at the upper right a five-cell brain derived from the medieval scholastic philosophers St. Thomas Aquinas and Albert Magnus.

For over two centuries philosophers, theologians and physicians argued over the location within the brain of the seat of the soul. Rather than an incremental process, with past discoveries providing the basis and inspiration for future efforts, progress was slow, uneven, and frequently regressive. The Greco-Roman physician Galen (A.D. 130–200) experimented with pigs, cutting the motor and sensory nerves. He speculated that the sensory nerves travel to the front part of the brain which, he held, was soft and able to receive impressions, whereas the motor nerves derived from the back part of the brain which was composed of a harder substance. Although Galen's speculations were wrong anatomically (actually the sensory and motor areas are just the opposite: motor in the front, sensory farther back), the attempt to localize "lower" motor and sensory functions to different parts of the brain served as a stimulus fourteen hundred years later for Descartes to formulate the first systematic account of the mind-brain relationship.

In his attempt to find the seat of the soul, Descartes set out with the assumption of two different created substances, body and soul (also termed "mind"). The body is extended and occupies space, while the mind is unextended and its essence is thought, pure thinking substance that may, but need not always, regulate the body.

> I must begin by observing the great difference between mind and body. Body is of its nature always double. When I consider the mind—that is, myself, insofar as I am merely a conscious being—I can distinguish no parts within myself; I understand myself to be a single and complete thing. Although the whole mind seems to be united to the whole body, yet when a foot or an arm or any other part of the body is cut off I am not aware that any subtraction has been made from the mind. Nor can the faculties of will, feeling, understanding and so on be called its parts; for it is one and the same mind that wills, feels, and understands. On the other hand, I cannot think of any corporeal or extended object without being readily able to divide it in thought and therefore conceiving of it as divisible. This would be enough to show me the total difference between mind and body, even if I did not sufficiently know this already.

But how can a spatial body affect or be affected by dimensionless, unextended mind? This was the "Cartesian impasse," the dilemma Descartes created for himself. In a letter dated 26 January 1640 addressed to a Father Mersenne, he set out his final solution to the mystery of the seat of the soul:

> This gland [pineal] is the principal seat of the soul and the place where all thoughts originate. The reason from which I derive this belief is that I find no part in all the brain, save this alone, which is not double. Now since we see only one thing with the two eyes, nor hear but one voice with the two ears, nor have but one thought at the same time, it must of necessity be that the different things that enter by the two eyes or two ears must go to unite in some part of the body there to be considered by the soul. Now it is impossible to find any other suitable place in the whole head but this gland. Further, it is situated the most suitably possible for this purpose, to wit, in the middle between the cavities.

By placing the soul within the pineal, Descartes believed he had resolved his impasse and clarified the philosophical relationship between mind and body. But this emphasis on the pineal said nothing about the relationship of mental activity to brain function. Nor was that his intention. Rather, Descartes's aim was to localize to an infinitely small area of contact, somewhat analogous to a mathematical point, the "higher" functions of the mind. While Descartes recognized the importance of the brain—indeed the substance itself and not the hollow ventricles—he had little interest in subdividing it. Mind interacted with brain via the tiny, centrally placed pineal and that was the end of it; what else was there to say?

In 1649 Descartes sent to press the manuscript of his last and greatest work, *les Passions de l'âme* ("The Passions of the Soul"). It contains his most detailed account of the interactions of mind and brain via the pineal. Six months later Descartes died of exposure after returning in the bitter cold from a 5 A.M. teaching session imposed by Sweden's Queen Christina. It was a sad end for so brilliant and complex a man. Although his localization of the mind to the pineal would not stand up to subsequent investigations, Descartes raised to a new and more elegant level the question of the relationship of the mind to the brain.

A contemporary of Descartes, the physician Thomas Willis did not accept the Cartesian emphasis on the pineal. He said that the cerebrum controlled voluntary actions and sensations and that the cerebellum, a structure at the back of the brain, was responsible for involuntary movements we would now ascribe to the autonomic nervous system, such as the actions of the heart and lungs. Although retaining the ancient divisions of the mind (sensation, imagination, and memory), Willis moved them from the ventricles to precise locations within the brain substance.

One hundred years after the deaths of Descartes and Willis, a French surgeon, La Peyronie, delivered a paper in Paris in which he described an experiment he had carried out on a young soldier who had suffered a severe head wound that extended to the depth of the corpus callosum, the fiber tract connecting the two cerebral hemispheres. When La Peyronie injected water into the wound, the patient would lose consciousness. With removal of the water the young man would wake up, inspiring La Peyronie to suggest that he had discovered the place where the soul "exerts its function." The paper soon achieved the status of a classic, at least among physicians, as proof for a primary seat of action of the soul. Today we recognize this claim as nonsense: Flooding the brain with water and then suctioning it out tells nothing about the location of mental powers, but merely disturbs the balance of fluids and electrolytes necessary for consciousness. But La Peyronie's paper was important for a different reason than as a clumsy and wrongheaded attempt to establish the location of the soul. It marked a shift from philosophy and theology toward medicine, toward a belief that mental faculties could be localized in specific areas of the brain.

But that transition from the spiritual to the medical was not smooth. In 1746, a physician of the French Guard, Julien Offray de La Mettrie, was fired from his post and his book *Histoire naturelle de l'âme* was condemned to be burned by the Paris parliament. La Mettrie aroused such enmity because he claimed that the brain was not just a passive receiver of impressions but the determiner of thought. After fleeing to Holland, he wrote a second book, *L'Homme-machine*, which elaborated upon the relationship of brain structure to thought. It was no accident, said La Mettrie, that the human brain, the most advanced of all known species, has the greatest number of convolutions in relation to body size. Next came the monkey, the beaver, the elephant, the dog, the fox, the cat, and so on. Such a view, which brushed aside all nonmaterial explanations for mental activity, upset the Dutch as much as his earlier book had affected the French. The Dutch government ordered *L'Homme-machine* destroyed, and in response La Mettrie fled once again, this time to Prussia at the invitation of Frederick II.

During the remainder of the eighteenth and into the nineteenth century, the debate about the relationship of the mind to the brain continued. A famous textbook of anatomy entitled "On the Organ of the Soul" stirred up the animus of no less a figure than Goethe, who wrote to its author that he should have confined himself to explaining how the nerve endings worked and left all discussion of the soul to those with the proper philosophical qualifications. His anger is easily understandable. To Goethe and many of his contemporaries, the soul pervaded all living things and could not be localized in one site. Besides, the brain offered opportunity and complexity enough without any need to introduce the ineffable.

"Before the labyrinth of the brain we are like those thieves who know all the alleys of Paris as well as the back of their hand, though ignorant of what's going on inside the houses," as one physician at the time put it. Knowledge of the happenings "inside the houses" came from a most unlikely source: an industrial accident.

* * *

On September 13, 1848, an explosion occurred near the small town of Cavendish, Vermont. Moments before, a team of railroad workers under the direction of their foreman, Phineas Gage, had poured gunpowder into a narrow hole they had drilled into a massive rock which blocked their way. A long iron rod was rammed into the hole by Gage for the purpose of tamping down the charge. By accident the rod scraped against stone, creating a spark which ignited the powder. The massive rod flew up with the force of a miniature missile and struck Gage just beneath his left eye. Although he suffered a severe head wound to the frontal lobes of his brain from which he was not expected to recover, Gage confounded his doctor and within a few weeks was walking about with no apparent injury. But Gage was a changed man, and the change was for the worse.

"His physical health is good and I am inclined to say he has recovered. . . . [but] . . . The equilibrium or balance, so to speak, between his intellectual faculties and animal propensities, seems to have been destroyed," wrote his doctor. "He is fitful, irreverent, indulging at times in the grossest profanity (which was not previously his custom), manifesting but little deference for his fellows, impatient of restraint or advice when it conflicts with his desires, at times pertinaciously obstinate, yet capricious and vacillating, devising many plans for future operation, which are no sooner arranged than they are abandoned." Gage's injury and the devastation it inflicted on his personality set off a furious debate in the latter part of the nineteenth century, a debate that continues today, between those who believe that all mental functions can, at least in principle, be localized to specific areas within the brain, and those who hold that the brain operates as a whole.

Among those favoring localization, none was more colorful or generated greater furor during his lifetime than Franz Josef Gall. As a child, Gall was impressed by apparent correlations between unusual talents in his friends and variations in the shapes of their faces and skulls. Large protruding eyes, for instance, seemed to be associated with superior memories. "I was forced to the idea that eyes so formed are the mark of an excellent memory. . . . Why should not the other faculties also have their visible external characteristics? From this time all the individuals who were distinguished by any quality or faculty became the object of my special attention, and of systematic study as to the form of the head."

Gall's obsession (and obsession it truly was) led him to conclude that the brain was composed of as many areas as there were faculties or feelings and, since these were located on the surface of the brain, they influenced the formation of the skull. Thus by skull palpation one could discover the moral, emotional, and intellectual faculties of the individual in question.

In a famous letter to Baron von Ritzer, Gall wrote of his intention to create a more faithful and truer portrait of humanity based on the brain. He proposed four assumptions: First, a person's moral qualities and intellectual faculties were innate. Second, their exercise depended upon brain organization and structure, in a word the organ's "morphology." Third, the brain acted as the organ for all inclinations and faculties. Finally, the brain was composed of as many particular organs as there are functions natural to humans.

Today only the first of Gall's assumptions would get him into trouble. In a culture such as ours where belief in political equality is firmly held, any suggestion that people may in fact be born with unequal abilities or intelligence meets with accusations from the thought police of "elitism" or even "racism." But in Gall's time all four assumptions proved equally offensive. Part of the problem stemmed from Gall's insistence on replacing philosophical speculation with observations drawn from nature. In doing this, Gall involved himself in a turf battle with philosophers who traditionally wrote and spoke about the mind. They deeply resented the intrusions into their ethereal domain by a lowly physician. The philosophers also resented Gall's overvaluation of people's innate inborn tendencies in comparison to the influence exerted on them by the environment and education. And this disagreement was no minor academic quibble. Gall was replacing the traditional emphasis on categories and concepts like perception, attention, judgment, imagination, knowledge, feeling, and will in favor of a method, based on nature, that emphasized the proportions, distributions, and mixtures of the "primitive forces" responsible for a person's uniqueness.

Gall's lectures were condemned by the Emperor Francis II, and in an 1801 imperial handbill Gall was accused of "materialism," at that time a serious breach of political correctness.

Eventually a commission of the Institut de France investigated Gall's claims. Their findings were not favorable. At this distance in time, it appears that the cards were stacked in advance, since no less a person than Napoleon was rumored to have imparted to the commissioners his own opinion that Gall was an "impostor" and phrenology nothing more than a fanciful theory.

Undeterred by such criticism, Gall defiantly traveled to France where he was welcomed by the Société de Médecine. Wealthy and influential patients sought consultations in order to learn more about themselves. The establishment of twenty-nine phrenological societies soon followed in the space of only a few years. Newspapers and magazines throughout Europe and in the United States featured major stories on phrenology along with cartoons and caricatures of phrenologists solemnly palpating heads and examining skulls. One of them in my personal possession depicts a gathering of fashionable guests at the conclusion of a formal dinner. The men are enthusiastically feeling the heads of their female companions. Everyone is laughing and seemingly having a good time except for one solitary guest who sits alone in a corner examining a skull.

This new "science" of mind and behavior appealed to both conservatives and liberals, hence its success. On the one hand it lent support to those non-egalitarians who found in phrenology an anatomical "proof" for innate biological differences, and hence the wisdom of the prevailing deeply exclusionary social arrangements. At the same time phrenology served as a vehicle for displacing traditional religions with a new ethic of self-improvement. One of the most popular and successful books of the nineteenth century, *Constitution of Man Considered in Relation to External Objects*, by George Combe, a Scottish lawyer turned phrenology zealot, extolled temperance, good hygiene, the value of the family, and the virtues associated with the work ethic. The book emphasized phrenology as a guide to self-knowledge and thereby the possibility for change. To know oneself and others phrenologically meant paying attention to dispositions, attitudes, interests, and aversions. Knowledge of these things could provide, it was believed, the basis for the education of children, the reform of criminals, and general improvement within the society.

Although the "science" espoused by the phrenologists now appears primitive and simplistic, the social orientation that it spawned sounds strikingly contemporary. In the United States, itinerant phrenologists traveled to small towns providing craniologic analysis and advice about the evils of drinking or smoking, the values of vegetarianism and hydrotherapy, the rights of minorities and immigrants. Enterprising phrenologists made available to an eager public various objects and inventions such as the Casket of Knowledge, an ornate box about the size of a short novel and containing forty cards along with a neoclassical bust in relief. The "most portable" of the phrenological systems, the Casket was both convenient and aesthetically pleasing. But the most ambitious phrenological invention was the "home for all," a house built in an octagonal shape (the closest to the perfection of the sphere) and constructed according to plans that even specified the choice of building materials. Thousands of these buildings, some of them still existing, were constructed along the upper Hudson River and in the Connecticut valley.

Gall's influence has persisted into the present day, but not in a way he would have anticipated or desired. In my office I have, as a kind of neurological curiosity from a simpler era, a phrenology chart. Across the top are the words, "Phrenology//The Proper Study of Man . . . 'Know thyself'." Depicted on the outer surface of a rather androgynous head are such scenes as a young boy standing over another youth he has just knocked to the ground. The numeral 6 identifying this scene directs attention to a code labeled "Names and Numbers location of the mental Faculties." Beside the number 6 on the list is written: "Combativeness—resistance, defense, criticism." Another faculty located toward the back of the head is depicted as Cupid complete with his bow. On the master list below the drawing is written: "Amativeness—connubial love, affection."

Today we know that there is no necessary correlation involving the size or shape of the skull, brain anatomy, or mental capacities. Surely Gall at some level must have known this too. Erasmus had a short small head and a very small brain, estimated as probably no more than 1160 grams in weight (the average male brain weighs between 1400 and 1600 grams). That is why he is always depicted wearing the specially constructed wig and large biretta that he favored to disguise his tiny head. Other examples of brilliant people with superior mental capacities but abnormally small head sizes include Anatole France and the German anatomist Wilhelm Roux, who was so troubled about the small size of his head that as a final deathbed wish he ordered that his brain be removed and destroyed at death so as to avoid its preservation as a medical curiosity. Such facts were available to Gall, and yet he persisted in palpating "bumps" on the skull and correlating them with mental faculties. Such is the nature of obsession. But, to be fair to Gall, he did not limit himself to intelligence but included other feelings and mental abilities. His importance today is based on his insistence that the brain, "a marvelous collection of apparatus," must form the underpinnings of any theory of mind or behavior. He emphasized observation, measurement, and biology in place of speculation or empty theorizing.

One of his beliefs has endured to the present day: the importance of the frontal lobes. Along with language (verbal memory as he referred to it), Gall assigned to the frontal lobes all intellectual faculties. He declared that a person's intelligence related to the extent of his or her frontal lobe development. This emphasis on the frontal lobes was to inspire investigations over the following two decades that would culminate in the discovery of a "speech center" in the frontal lobe of the left hemisphere. Although the particulars of phrenology turned out to be wrong, or at the least highly exaggerated, Gall pioneered one of the two major theories about the organization of the brain, localizationism.

In opposition to Gall, his pseudoscience of phrenology, and localization of brain function in general, was Marie-Jean-Pierre Flourens, a child prodigy who enrolled in medical school when only 15 years old, had himself appointed perpetual secretary of the Academy of Sciences at 39, and at 46 bested Victor Hugo for a place in the French Academy. Flourens's genius combined finely honed surgical skills with an insatiable curiosity about how the brain is organized. While others only speculated or, like Gall, studied the brain at a distance, Flourens completely uncovered the brains of a host of animals, including dogs, pigeons, and rabbits. After isolating various brain parts, he then either removed or destroyed them. He observed that destruction of the cerebral lobes (the most anterior part of the brain, which in mammals contains the cerebral cortex) produced a loss of spontaneity. The animal would sit immobile, although still capable of movement if the examiner stimulated it. Flourens then cut the cortex away in successive slices and observed the effects on the unfortunate creatures' behavior (obviously, the concept of animal rights was a long way off). He found that it was the amount of cerebral tissue removed rather than its location that determined the final effect. From these grisly experiments Flourens arrived at some rather weird conclusions. The cerebral hemispheres were not the immediate cause of muscular movement nor the site that coordinated movement, but rather, in his view, the areas of the brain responsible for "volition" or the "self will" of the animal. While Flourens was willing to admit that elementary sensations and simple reflex movements could be localized within the brain, he maintained that the "higher" brain functions like perception, intellect, and will are unified and operate together with the entire cerebrum as their "exclusive seat."

Flourens was a religious man who held strong beliefs in a unitary soul and an indivisible mind. In 1846 he published a book (dedicated to Descartes) accusing Gall and his colleagues of undermining the unity of the soul, immortality, free will, and the existence of God. A religious orientation is so rarely encountered in contemporary neuroscientists (at least those known to me) that it is all too easy to ignore the powerful influence of these beliefs on Flourens's theories about the relationship of mind and brain. Earlier we spoke of the temple as serving as a metaphor for the human brain. Here the appropriate metaphor seems more likely a cathedral, or more specifically, the awe one might feel when visiting a cathedral. The unwillingness of Flourens and his followers to think of the cerebral hemispheres as like the rest of the brain was an expression of their belief that, in order for us to retain our special identity as humans as opposed to animals (Darwin's *Origin of Species* would not come for another decade), some part of the brain must remain sacrosanct. For Flourens and his followers in early nineteenth-century Europe, the cerebral hemispheres performed that function.

Over the next quarter century the debate continued between the localizationists, represented by Gall, and supporters of Flourens's "equipotential" theory. In 1861 the pendulum swung decisively in the direction of the localizationists. In that year Paul Broca, a physician with a life-long interest in physical anthropology and a founding member of the Société d'anthropologie, reported on his examination of the brain of a patient he had examined six days before his death. The nickname, "Tan," was given the patient in the hospital because this was the only word he could say. At autopsy, Broca found an injury to the left frontal lobe. This discovery of a localized area responsible for speech confirmed Gall's earlier assignment of speech to the frontal lobes. But it also created a dilemma: Why would speech be limited to just the left frontal lobe instead of *both* frontal lobes? Although we are comfortable now with the

concept that the two sides of the brain perform different functions,*
Broca and his contemporaries knew nothing of hemisphere specialization and believed in the functional symmetry of all bilateral organs.

Over the next half century, others discovered additional areas concerned with speech, vision, orienting oneself in space, even an area
concerned with the sequences involved in dressing. The inner sanctum
of the cerebral hemispheres had been broached. From then on no part
of the brain would be considered unapproachable.

But the study of anatomy and structure can only reveal so much. What
are the principles of operation within the brain? How—to put it at its
most basic—does the organ work? Anatomical dissections and theorizing about one's findings can never answer these questions. If the brain is
to be comprehended, an interest in brain structure must give way at
some point to neurophysiology, the study of brain function. The earliest
efforts employed electrical currents applied to the cerebral hemispheres.

Stimulation of the hemispheres of a dog's brain was found to elicit
movements of its legs and paws on the opposite side. Destruction of
those same areas of the brain results in permanent paralysis of the
limbs. Thus, contrary to a long-standing belief, the hemispheres do par-
ticipate in movement.

If mind emerges from the activity of the brain, how are thoughts,
feelings, and other mental activities organized within the brain? The
nineteenth-century philosopher Herbert Spencer—the last major
philosopher to make an original contribution to our understanding of
the brain—suggested that mental development occurs along a continuum, with "higher mental processes" evolving from simple reflexes and
instincts. According to Spencer, there is no reason for drawing a sharp
distinction between mind and brain. Nor is there any reason to assume
that the cerebrum, enlarged as it is in humans compared to other creatures, functions according to principles any different from those governing activity elsewhere in the brain.

Spencer exerted great influence on John Hughlings Jackson, an
amateur philosopher turned neurologist. Of a brooding, melancholy dis-

---

* A *New Yorker* cartoon shows a business executive flanked on his right by a bearded, beaded hippie and on his left a prim young man in a business suit resembling a
character from *The Return of the Nerds*. "These are my right brain and my left brain,"
says the executive.

position, Jackson was a keen observer of the effects of injury and illness-es of the brain and thought a great deal about the implications of what he observed. He maintained a particular interest in epilepsy, which he considered the key to the functions of the brain. One form, which bears his name, was of particular interest to him. In focal epilepsy, the seizure will progress in a definite order or "Jacksonian march" involving usually the face first, then the hand and arm, then the leg on the same side, and finally a generalized seizure that catapults the patient to the ground. During one of these seizures the patient often experiences a dis-turbing and inexplicable temporary loss of the "will" to move. Even more frightening and disturbing are experiences of an unexplainable fear or dread. At such times the patient feels distant from himself, strangely altered, the world and things around him threatening and ominous. All actions are temporarily impossible, and the "higher" men-tal functions of reason, memory, and emotions are altered and beyond conscious control. With the end of the seizure, the patient returns to his or her normal state.

On the basis of his observations of patients undergoing these "psy-chic" forms of epilepsy, Jackson concluded that all mental functions, from the most automatic to the most seemingly ethereal, are dependent on the workings of sensory-motor circuits within the brain. This must be so, he concluded, because "In cerebral damage the parts suffer as much in voluntary operation as they do in automatic operation, in other words, the will is affected."

At the end of a paper, "The study of convulsions," he had this to say: "It is asserted by some that the cerebrum is the organ of mind, and that it is not a motor organ. Some think the cerebrum is to be likened to an instrumentalist, and the motor centers to the instrument—one part is for ideas, and the other part for movements. It may, then, be asked, How can discharge of part of a *mental* organ produce *motor* symptoms only? . . . But of what *substance* can the organ of mind be composed, unless of processes representing movements and impressions? . . . Are we to believe that the hemisphere is built on a plan fundamentally dif-ferent from that of the motor tract? . . . Surely the conclusion is irre-sistible that 'mental' symptoms . . . must all be due to lack, or to disorderly development, of sensori-motor processes."

Jackson's achievement was not that he localized mental functions to the brain—an assignment already established by this time in the late

1870s—but that he offered the intriguing suggestion that the workings of the mind consisted of motor responses to sensory stimuli. While the complexity of these motor responses might vary from the automatic kick of a knee in response to the tap of a reflex hammer to the formulation of a scientific theory, to the composition of a sonnet, each instance involved balancing forces at work to control a gradation of movements "from the most voluntary to the most involuntary."

Jackson envisioned brain organization to be based on a pyramid or hierarchy of increasingly complex functions. Under ordinary conditions, the "higher" functions of rational thought and conscious intention, located in the cerebral hemispheres, especially the prefrontal lobes, hold in check the more instinctual and automatic responses of the "lower levels" located beneath the hemispheres and within the brain stem and spinal cord. His thinking was a blend of evolution and class consciousness appropriate for an age deeply influenced and yet haunted by revolutions and insurrections. Jackson spoke of "control" and "release," "dissolution" and the consequent "welling up" of "primitive brain levels" after injury to "higher" centers. Indeed brain injury, whether due to stroke or tumor or by other means, involved a reversal of the evolutionary process and the ascendence of lower, more "atavistic" forces. On occasion Jackson even employed political language to describe his theory. "If the governing body of this country were destroyed suddenly, we should have two causes of lamentation: 1. the loss of services of eminent men; 2. the anarchy of the now uncontrolled people." Experts in the social sciences incorporated Jackson's neurologic hierarchy. The French social psychologist Gustav Le Bon, who studied the behavior of people in crowds, spoke of a mob as "spinal man," an organism powered only by reflex and completely separated from the higher centers of reason and judgment.

Sigmund Freud incorporated Jackson's hierarchical organization in his theories about mental illness. To speak of "regression" to earlier and more primitive behaviors is only to employ neurological terminology in the service of psychiatry. In describing the relationship of the conscious rational ego to the instinctive, unstructured id, Freud spoke of "a man on horse-back, who has to hold in check the superior strength of the passion-driven horse."

Note that in the space of one generation, theories about the brain moved from analogies drawn from theology and philosophy to analogies drawn from politics and social science. Indeed, the emerging under-standing of the brain depended very much upon the philosophical-reli-gious persuasions of the investigators. While Flourens, as mentioned before, was deeply religious, Broca supported left-wing anticlerical activity in France aimed at replacing the ethereal philosophy of Descartes with a more down-to-earth materialistic orientation. What better way of doing this than by proving that speech, mankind's most distinguishing ability, could be physically located within a specific point in the brain? That way the soul, or at least one of its powers, was pinned down in space. As we shall see, each generation's thinking about the brain borrows from the social, political, and technological climate of the period. At the moment, computers and other emerging technologies are thought to provide clues, if not outright explanations, of how the brain operates. (We will have more to say about this later.) This infatua-tion with technology is both a help and a hindrance: One generation's awe at new and powerful technologies only draws yawns decades later after the development of yet newer and more powerful technologies. Besides, there is a bootstrap problem here, a conundrum which, I must admit, tempers my enthusiasm that any fully satisfactory explanation for the brain will ever be forthcoming. In my slide collection I have an image that expresses my concern:

A man sits on a chair while pulling at a ball of twine. As the strand passes through his fingers, he studies it closely, as if searching for some hidden meaning. Only the observer of this scene can see that the man's task is impossible because his head is the ball of twine from which the strand originates. A similar self-referential paradox underlies our under-standing of the brain and its relationship to what we refer to as the mind. We employ our brains in order to understand how the mind is related to the brain from which everything emerges, including our men-tal efforts toward achieving this understanding. Is this the correlate of the oriental maxim that the eye that sees cannot see itself? Is the process something like biting one's own teeth? I think we face here a variant of Pascal's Wager. That philosopher declared that a belief in God makes sense because if God does not exist, everything ends with death and is essentially meaningless. But if God exists and we believe in Him, we will presumably receive whatever reward awaits believers. I feel that way about the brain. Even if we cannot understand it completely, we can sure learn a lot of useful things about it and enrich our lives in the process. And if I am wrong and we really can understand all aspects of its functioning, what greater goal could there be, considering that our entire mental life originates with the brain?

Notice that John Hughlings Jackson prompted in me these philosophical reflections. And how appropriate. Jackson more than anyone we have discussed so far formulated an original theory of the brain, a theory that persists today and continues to influence neurology, psychiatry, and theoretical neuroscience. Moreover, he did this from an orientation that was both scientific and philosophical.

Jackson's hierarchical theory initiated another breakthrough: recognition that ordered mental processes could occur outside of conscious awareness. With this discovery came a new dilemma: Is the self a unity, as supposed by previous thinkers, or do its various patterns of "dissolution" imply a multiplicity? This is a conundrum we will take up throughout this book and is the essence of our inquiry. But first some necessary groundwork must be established. Before we speak of the elusive nature of consciousness, mind, and identity, we must start with what initially might appear as a simpler subject: vision. But the brain processes underlying vision are far from simple. Indeed, the seeming simplicity and unity of our perceived visual world is an illusion, a construct of our brain's organization. Moreover, discovery of the principles of visual organization within the brain provided neuroscientists with one of their first proofs of the brain's modular organization.

CHAPTER TWO

Doctor Mountcastle's Curiosity

Vision is our most important and valued sense. Ask someone which of their senses they would least be willing to lose and you won't have long to wait for your answer. Blindness is universally frightening because we are first and foremost visual creatures. Our brain is responsible for this preference for things visual since the organ has more nerve cells devoted to vision than to sound, touch, smell, or touch. Vision also provides us with a more unified perception than any of the other senses. Close your eyes while sitting on a park bench and try by means of the other senses to gather a completely satisfying conception of what's going on around you. Only vision can do that. Your other senses provide only partial information; vision is the final arbiter. Yet the process by which the eye and brain inform us of what's going on in the world is far from simple or straightforward.

When someone looks at something, the light reflected from the object produces a miniaturized image of the object on the retina: the clustering of photoreceptors and connected nerve cells at the back of the eye (see diagram 4). The retina transforms the image into impulses which travel along the cablelike optic nerve to the visual area at the back of the brain (V1). Total blindness results from the severing or destruction of the optic nerve fibers anywhere along their path from the retina to V1. Complete destruction of V1 also results in blindness. Partial damage produces varying types of visual disorder.

Two decades of Nobel prizewinning work by David Hubel and Torsen Wiesel at Harvard revealed that vision depends on two kinds of neurons in V1. One is sensitive to line orientation and the other to wavelength. Further, the wavelength-sensitive neurons influence each other's responses. Some fire in response to what we see as a green light and switch off in response to what we see as a red light, or vice versa, while

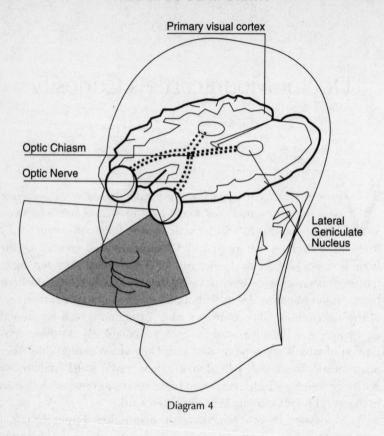

Diagram 4

others react similarly to yellow and blue. Surrounding V1 is a second visual area, V2, which also contains neurons sensitive either to line orientation or to wavelength.

But seeing involves far more than just lines and colors. Illumination changes from moment to moment; the wavelengths of reflected light from the surfaces of objects vary enormously; retinal images are never the same from second to second—yet despite these variables, the scene before us retains a constancy and coherence over time. So where in the brain does the neural processing take place that is responsible for color constancy, movement, and the other components of the visual experience?

One way of answering that question involves observing the effects on vision produced by brain damage secondary to strokes, tumors, or other

26

brain afflictions. Although this may be helpful in individual instances, most naturally occurring brain disorders show little regard for anatomical subtleties and involve large and varied areas. In order to get around these vagaries, neuroscientists employ highly invasive experiments with animals that would be impossible to do in humans. For instance, Hubel and Wiesel learned about the line- and wavelength-sensitive neurons by inserting tiny electrodes into the brains of anesthetized animals. The monkey is a favored experimental subject because of its close kinship to humans. As a result we know more about the visual system of the monkey than we do about our own. We know the monkey brain contains visual cells sensitive to color, lines of particular orientations, patterns of moving dots, even faces. Since you may be wondering how neuroscientists arrived at these conclusions, let me provide a brief description of two of the experiments.

While probing an area of the macaque monkey's brain known as V4, Semir Zeki, a neurobiologist at University College in London who has devoted a lifetime to research on functional specialization in the visual cortex, discovered neurons that fired only when the animal was exposed to a pure color stimulus devoid of any recognizable form (a Mondrian painting, for instance). In addition, these cells continued to respond under different lighting conditions. This behavior at the cellular level mirrors our own visual experience where we recognize colors under widely varying lighting conditions. Moving the electrode into a different area, V5, reveals neurons that respond to moving patterns of black and white squares.

In humans, a PET scan replaces intrusive electrodes as the means of studying the brain's visual areas. PET, an abbreviation for Positron Emission Tomography, uses a cyclotron, a computer, and a scanner. The cyclotron prepares chemicals labeled with trace amounts of a radioactive element. For imaging the brain, the patient is given by intravenous injection in the arm radioactively labeled sugar—the primary nutrient of the brain. The sugar then passes through the bloodstream into the brain where the PET scan camera records the radiation emitted from the labeled sugar. It then produces images showing the rate at which each brain area is consuming the sugar. The more active a particular brain area, the more sugar it uses and the more radiation recorded. "A glorified Geiger counter" is how one neuroscientist describes the PET scanner.

PET studies of normal volunteer subjects reveal that the brain areas for visual coding of color and motion correspond to the V4 color center and the V5 motion center identified in Zeki's earlier experiments with macaques.

Based on PET scan findings and the work of Zeki and others, neuroscientists now believe our visual experience is built upon the "division of labor" principle. Zeki suggests that, rather than a photographic plate and its interpretation (the oft-repeated "classical" metaphor for the visual process), vision is more correctly compared to a post office in which different visual "messages" are parceled out to their appropriate areas.

Nerve impulses coding different types of information arrive at V1 and are distributed anteriorly to more specialized processing areas (V2, V3, V4, V5) toward the front of the brain in what used to be referred to as the visual association areas, or "the entering place of the visual radiation into the organ of the psyche," as one nineteenth-century neurologist with a poetic bent put it. V2 is thought to be a general facilitator of visual information flow; V3 deals with recognition of movement involving recognizable forms; V4 with color; and V5 with movement in general.

Normally these different modular functions are unified into a seamless whole. When I look out my window and see a new Mercedes-Benz driven down the street, color (the car is blue) is processed separately from high-resolution form perception (it's a Mercedes) and movement and stereoscopic depth (the car is moving at approximately thirty miles an hour and against the background of the houses across the street). So far no one has come up with an explanation of why my visual image is unified despite the fact that the separate areas are operating independently of each other within the brain.

And damage to these different areas results in highly distinctive disturbances.

If V4, the area mediating color, is injured, the patient sees the world only in shades of gray. But more is involved here than simple color blindness. Achromatopsia, the neurologic term for the disorder, not only robs the victim of color in the present and future (the condition is incurable) but affects the person's memory for color as well. Form, depth, and motion appreciation are unaffected. The experience is not like losing one's memory for colors but is more like an existential loss in which the very concept of color has irretrievably disappeared.

A lesion in area V5 results in "motion blindness" or akinetopsia, the loss of the appreciation of movement. Only objects at rest can be seen; anything that moves disappears. A woman affected with this disorder reported that she experienced the world as a series of static images. As a result, she had great difficulty with everyday activities like crossing streets or pouring drinks. Cars did not seem to be moving but instead changed positions so that a car that appeared far away suddenly loomed in front of her. Tea when poured from a cup didn't appear to flow but formed a solid arc.

Damage to the prestriate areas that spares V1 leaves intact the ability to recognize the separate components of form (shapes, angles) while depriving the person of *meaning*. The victim evidences a kind of "mind blindness," or visual agnosia, the loss of the ability to identify what is seen. One patient with such a disability drew a perfectly acceptable copy of St. Paul's Cathedral but had no idea what he had drawn. Other varieties of visual loss include the loss of form and the resultant inability to identify or draw simple shapes, or the retention of only single color sensitivity so that all green objects are identified as grass. A particularly intriguing patient described by Jules Davidoff, a neuropsychologist at the University of Essex who specializes in color vision, lost the ability to identify common objects with their colors. When asked to color in a drawing of a carrot, he filled in the outlines in olive. At that point he said out loud, "What color is a carrot?" He then said "orange," looked at the olive carrot and said, "That is not orange, so it must be wrong. But it doesn't look wrong." This patient's strange performance suggests to Davidoff that the brain has separate storage areas for color and object names and even one for associating objects with their colors.

My point here is not to provide an all-encompassing description of the various ways vision can break down, but to emphasize how what originally seemed a unified process actually consists of a division of labor. Each separate module—form, color, movement—contributes to the unity of visual experience. And this will vary from one creature to another. Thus when a monkey confronts a lion, one group of visual cells in the monkey's brain may respond to the bearded mane, another to the contour of the head and its attachment to the body, and yet another to the sticklike elements that compose the legs. Similar super-recognition cells exist in other animals, according to the unique conditions of their existence. Sheep, for instance, possess visual brain cells that respond to another sheep or to potential predators like wolves or people. Activation of any one of these groups of cells may be sufficient to bring into action all of the others. Each of these components operates separately and can suffer breakdown independently of the others. Indeed, only when vision breaks down in some way do we obtain insight into the separate organizational components.

Consider the experience of Maureen, a fifty-seven-year-old interior decorator. Described by her family as a vibrant and talented woman, Maureen complained one day of a "novocainelike" numbness and an odd feeling in her hands, as if continuously held in an abnormal position that she likened to the head and neck of a chicken. When asked to explain this, she flexed her wrist and extended and brought into opposition the thumb and four fingertips of each hand. Over a two-week period, her relatives noted just such a peculiar positioning of Maureen's hands. She also started to lose her balance easily, became extremely forgetful, and, most puzzling of all, she would move her head from side to side whenever she was shown something and asked to name it. She had no difficulty in recognizing something shown her if someone verbally described it or she was permitted an opportunity to touch it. If asked to describe a scene or even something as forthright as the food and objects on her meal tray, she could occasionally name one object which she then repeated for everything else on the tray, much to her own dismay and the dismay of the doctors caring for her. By the end of the second week in the hospital she began having intermittent jerking movements of her whole body and even greater difficulty scanning and describing objects and pictures. She talked less and less, often speaking to invisible persons in her room. Nine weeks after entering the hospital, she died. At autopsy Maureen's brain showed widespread loss of nerve cells throughout her brain, especially toward the back near the visual area and its connections. This illness, fortunately rare, is known as Creutzfeldt-Jakob disease (CJD), and the variety that killed Maureen shows a predilection for the visual and visual association fibers. As with Maureen, the afflicted person typically can see quite clearly but cannot identify what they see: visual agnosia. The recognition defect is specific for vision since recognition remains intact for touch or sound.

During our exploration of the modular brain we will encounter other people like Maureen who suffer from various kinds of agnosia. My point in introducing the topic here is to illustrate how the visual experience consists not only of color, motion, and other "elementary" modules that we usually think of when considering vision, but also includes psychological elements as well. In the event of brain injury one or more of these modules may malfunction.

Semir Zeki, from whom I have borrowed several of the above examples, has concluded after a lifetime studying vision that these visual areas connect with each other via "reentrant connections which allow information to flow both ways between different areas." But how and where does all of this separate visual information get unified?

The most intuitively appealing explanation would be that all the specialized areas communicate their contributions to one central organizing area (our General Manager raising his head again) where synthesis takes place. But Zeki is convinced that in this instance our intuition is wrong both philosophically and scientifically:

> Philosophically, that solution begs the question, because one must then ask who or what looks at the composite image, and how does it do so. That problem is beside the point, however, because the anatomic evidence shows no single area to which all antecedent areas exclusively connect. Instead the specialized areas connect with one another, either directly or through other areas.

Moreover—and this is the main point—all of this takes place in the absence of a central processor.

But Zeki reaches another equally profound conclusion. Since he stated it so clearly in a *Scientific American* article, let me quote him directly:

> My colleagues and I have developed a theory of multistage integration. It hypothesizes that integration does not occur in a single step through a convergence of output onto a master area, nor does integration have to be postponed until all the visual areas have completed their individual operations. Instead the integration of visual information is a process in which *perception and comprehension of the visible world occur simultaneously.*

I placed the last portion of Zeki's comment in italics for a reason. It is nothing less than a concept-shattering notion: We don't perceive something, think about it or otherwise process it within the brain, and then comprehend it; but, rather, perception and comprehension occur simultaneously. This notion does away with all of the traditional metaphors (General Managers, homunculi, Master Centers, specialized brain areas communicating information in tandem, and so on) in favor of a process involving separate modules interacting with each other without any central control station. On occasion these modules can exert a sufficiently powerful influence to override our most firmly held beliefs. For instance, optical illusions do not disappear when an explanation is provided about how they are constructed. Paintings and drawings with strong perspective due to converging lines or gradients of texture can lose the depth sensation if the images are presented in equiluminant colors. An illusion like the famous vase/faces will also be seen differently depending on difference in luminance between foreground and background. (See drawings below.) With a high luminance-contrast image (black against white), one sees either the faces or the vase but not both together. At equiluminance, the two perceptions flip rapidly back and forth and with some practice both vase and faces can be seen simultaneously. This is because with equiluminance, the division between figure and ground becomes less distinct or may even disappear altogether.

Does the modular system hold just for vision? Or is modular organization basic to the brain's operation as a whole? To answer that question it's necessary to say a few words about the organization of neurons in the brain.

Two basic neuronal patterns exist in the cerebral cortex. A horizontal pattern contains six individual layers, each with its own predominant type and shape of nerve cell. At right angles to the surface of the cortex is a second pattern of neurons extending vertically from the surface down through all the layers. Electrical recordings made in the late 1950s by Vernon Mountcastle of Johns Hopkins University revealed that all neurons within a particular region of cortex no more than a few millimeters square shared overlapping receptive fields; i.e., sensory stimulation delivered to a specific area of the body activated the neurons. In one cortical area all neurons might have receptive fields involving a single finger, while another area may have a receptive field for the tongue or the arm. However, the neurons within a particular receptive field responded to different sensory stimulation. Some fired in response to a light touch on the skin, others to a pinprick or deep pressure. Moreover, all of the neurons

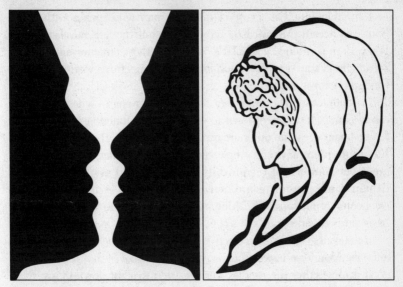

The vase/faces illusion   The wife/mistress figure

encountered by the electrode as it penetrated downward from the surface in a straight line responded to the same stimulus. But if the electrode entered the brain at an angle, so that it penetrated more nearly parallel rather than perpendicular to the surface, it encountered neurons that responded to stimulation of a different type. Thus a neuron responsive to light touch might have beside it a neuron sensitive to deep pressure in the same body part. Mountcastle described these vertically oriented clusters of functionally related cells as "columns." In time the operative word would change to "modules." But whatever term is employed to describe them, they are defined in three general ways: the nature of the stimulus that activates them; their behavioral response to activation; and their connection patterns with all other parts of the brain.

On 21 April 1993 I met with Vernon Mountcastle, who at 75 years of age was then perhaps the most eminent and revered neuroscientist in the United States.

"The brain is a complex of widely and reciprocally interconnected systems. The dynamic interplay of neural activity within and between these systems is the very essence of brain function," Mountcastle told me as we spoke in his office at The Zanvyl Krieger Mind/Brain Institute at Johns Hopkins University. A tall lean man who speaks with a soft Southern accent, Mountcastle reminded me of the late novelist Walker Percy. Both combined a scholar's curiosity with a gentleness and reserve seemingly ill suited to the unforgiving and competitive worlds of literature and neuroscience.

To Mountcastle the brain is a "distributed system" where the command cannot be pinned down to one area, but function resides in different brain areas at different times depending on the circumstances. Rather than employing a single central control center, the brain operates as a complex of reciprocally interconnected systems, and "the dynamic interplay of neural activity within these systems is the very essence of brain function." Mountcastle has come to this view on the basis of his experiments.

In the course of his microelectrode insertions into heavily anesthetized animals, Mountcastle was most surprised to discover that it mattered a good deal whether the needles were inserted straight down in a vertical pathway or at an angle somewhat parallel to the surface of the cortex. With vertical penetration all of the neurons responded to the same stimulus. When the electrode path deviated from the vertical and tended towards a more elliptical approach, the neurons encountered along the way responded to different sensory stimuli.

"The general idea is as follows: the large areas of the brain are themselves composed of replicated local neural circuits, modules, which vary in cell number, intrinsic connections, and processing mode from one brain area to another but are basically similar within any area. Each module is a local neural circuit." Multiply the number of modules within the brain by the number of nerve cells in an individual column, and you have a figure that corresponds to the minimum number of functionly important cells in the human brain. In Mountcastle's original estimate, he put the numbers as 400 million minicolumns and more than 44 billion neurons, but he believes now that the true figures are even higher. But the important point is not the number of modules, or the number of cells making up a module, but that the brain's incredibly complicated operation can be understood only in terms of the interaction of many similar repeating units.

"This suggests that the neocortex is everywhere functionally much more uniform than hitherto supposed. Further, its avalanching enlargement in mammals, particularly in primates, has been accomplished by the replication of a basic neural module, without the appearance of new neuron types or of different modes of internal organization," says Mountcastle.

One of the implications of Mountcastle's findings is that the modular organization of the brain provides yet another support for the idea that the uniqueness of the human brain does not derive from any special principle of organization. Further, the modules are not fixed and immutable but reorganize themselves on the basis of experience. For instance, some modules may be concerned with sensation from one of the fingers. If that finger is surgically removed the modules normally receiving impulses from the finger grow silent. But over the next several weeks the modular arrangement of neurons throughout that deprived cortex begin to respond to stimuli coming from the other fingers. Likewise, if the surgical lesion involves not the finger but the part of the cortex that normally receives sensory impulses from the finger, the job of representing that finger falls to neurons in an adjacent cortex which normally represent the other fingers. Such experiments point to the dynamic nature of the brain's modular organization.

The brain's response to injury provides one proof for this, according to Mountcastle. "The recovery of function after injury is a slow and gradual process that rarely reaches full completion. This is just what you would expect in response to an injury involving a small number of nodes in a multimodal system."

From the beginning of his research more than half a century ago, Mountcastle has always been wary of theories that emphasize hierarchical arrangements within the brain. "Most people don't think hierarchically anymore; they shy away from saying, 'This function resides *here*.' Instead we now believe the brain is arranged according to a distributed system composed of large numbers of modular elements linked together. That means the information flow through such a system may follow a number of different pathways, and the dominance of one path or another is a dynamic and changing property of the system."

Neuroscientists have turned up modules within the brain that map such things as visual space, body surface, and tone frequency (mediated by different locations within the cochlea of the inner ear). The organization of these maps depends upon the establishment and maintenance of connections between the sensory organs and interlocking modular centers within the brain. Alterations anywhere along this path alter the modular organization. Functions previously carried out by the altered area are transferred by modular reorganization to a nearby area. The younger, less developed the brain, the more efficiently do the modules reorganize themselves. At the moment the mechanisms of modular reorganization are not clear, but in the young they most likely involve regrowth and movement of neurons. In adults the mechanism is thought to be different since nerve cell regeneration is unusual. Instead, most neuroscientists believe that reorganization involves changes in the effectiveness of existing but formerly unused connections. Modification of modular maps take place when we learn new skills or enhance our perceptual abilities in new ways, such as learning to paint or play the piano. But this, of course, is a long way from proving that our sensations, emotions, memories, and thoughts—our most personal mental activities—are the result of the parallel operation of modules throughout our brain. In order to prove that we must show how these mental functions are encoded in the brain and how they are altered in response to injury or damage. In our exploration of these matters, we will discover that many of these operations aren't at all what we would imagine them to be or how we might have thought to design them.

CHAPTER THREE

"A Power of Will"

J ohn Hughlings Jackson provided the pivotal insight that even the simplest of actions requires the coordination of millions of neurons working together toward the expression of our will. Indeed purposeful action and movement are the "purpose" of all brain activity. We make up our mind to do something and, all things being equal, we *just do it*. Or so we like to think of the process. But the relationship between brain activity and the expression of our will is not only stranger than we think but perhaps even stranger than we *can* think.

While on a train the physicist Ernst Mach collapsed, the whole right side of his body paralyzed. Over the next several weeks his weakness began to remit. As it did, Mach noticed that his unsuccessful efforts to move his arm and leg were not accompanied by any sense of fatigue or strain. But with additional improvement, things changed; each small gain in the ability to move the limbs brought with it a sense of heaviness and resistance, as if the arm and leg were being held down by enormous weights. The harder he tried to move, the greater his fatigue and the greater the effort required. This inner sense of exerting a mental force against a feeling of inner resistance was later described by an anatomist, Alf Brodal, after his stroke: "Subjectively, [this] is experienced as a kind of mental force, a power of will. . . . It is as if the muscle was unwilling to contract, and as if there was a resistance which could be overcome by very strong voluntary innervation. . . . This force of innervation is obviously some kind of *mental* energy which cannot be quantified or defined more closely."

"A power of will"—a strange designation when you think about it. Who is doing the willing? What is the cause of the inner resistance? And the origin of the sense of effort? Certainly one need not suffer a stroke or other brain disturbance in order to experience this sense of pushing against inner resistance. Getting out of bed early on a weekend and forcing oneself to the fitness center; spending time with a person or a project that bores—at such moments a kind of mental inertia or fatigue exists that must be actively overcome, sometimes by what seems a superhuman effort.

The exercise of will traditionally occupies the further end of a continuum working upward from totally involuntary actions. A sneeze occurs spontaneously while a decision to change careers involves a sometimes agonizing voluntary exercise of the will. But obviously such a continuum from the involuntary to the deliberately willed is an oversimplification. Most of us can manage to stifle that sneeze if the occasion is august enough; and many decisions involving careers and other weighty matters are sometimes entered into impulsively. William James attempted a clarification which is only partially helpful: "Desire, wish, will, are states of mind which everyone knows, and which no definition can make plainer. We desire to feel, to have, to do, all sorts of things which at the moment are not felt, had, or done."

But we can bring about changes without seeming to exercise any deliberate act of will. If you fall asleep with your arm twisted into an awkward position, it's likely that eventually you'll turn so as to release pressure on the limb. In this case the "decision" to turn occurred outside your awareness, a precondition for willed action. One thing at least seems certain: " . . . we may start with the proposition that the only *direct* outward effects of our will are bodily movements" (William James, again).

Attempts at equating will with bodily movement started with the late eighteenth-century French philosopher Pierre-François Gonthier de Biran. Although he failed to achieve his lifetime goal of writing an all encompassing work with the grandiose title *The Science of Man*, Biran first emphasized the linkage between will and movement. "If the individual did not *will*, or were not determined to begin to move himself, he would know nothing. . . . He would not have any sort of existence, he would not even have the *idea* of his own self." This will-movement linkage made a good deal of sense to Biran's contemporaries. La Mettrie, whom we met in Chapter One, considered the mind a "thinking and feeling machine," and what characterizes a machine more than some kind of movement? Such reasoning led to an enthusiasm for automata, cleverly constructed machines contrived to simulate biological actions. The most ingenious, constructed by J. D. Vaucanson, included a duck which paddled itself about, and a tiny flutist who played several popular tunes of the period. Attempts to create biological automata soon followed.

An early and overly zealous brain researcher, F. L. Goltz, in an experiment straight out of a horror movie destroyed a dog's brain by means of a jet of water injected under high pressure through a hole in the skull. Goltz and the animal toured together in a kind of macabre road show to various medical meetings. When placed on the demonstration platform, the dog made no movements and appeared to be asleep. Noise would wake it up, a painful pinch would cause it to growl. When placed upright, it would walk in a mechanical way and would swallow food put in the back of its mouth. Otherwise the dog didn't act much like a dog, took no notice of its surroundings, and, most importantly, carried out no actions that could be even remotely considered voluntary.

Goltz's grisly experiment proved that fairly complicated actions can take place without benefit of a brain and strictly on a spinal cord level. There was no reason to doubt that a similar division between involuntary and voluntary actions existed in human beings who, as La Mettrie had claimed all along, are a kind of machine. But if involuntary actions could be handled very well at the level of the spinal cord, what part of the brain was responsible for *willed* action? The cerebrum, particularly the outer margin of the cerebral cortex, seemed the most likely site. Another Goltz-inspired search-and-destroy mission targeted the part of the cortex responsible for movement of the opposite side of the body. Rather than leaving the animal permanently paralyzed, the lesion's effect was only temporary. Soon "machinelike" movements reappeared. What remained absent, according to Goltz, was the "ability to initiate an action, in a smooth and sure fashion, of precisely those groups of muscles whose movements would be necessary to reach a goal, to achieve an end that was desired in a certain situation."

Today neurologists speak of a "duality of motor function" when describing the distinction between willed and involuntary action. A patient after a stroke, for instance, may be paralyzed on the side of the face opposite the brain damage. When asked to show his teeth or grimace, the face on that side will not move. But if a joke is told which makes the patient laugh, the paralysis will temporarily disappear. The dissociation here is between willed facial movement controlled by the cerebral hemisphere, and facial movements directed by brain areas beneath the cerebrum, especially parts of the limbic system which are known to be involved in the experience and expression of emotion. Another patient will not be able to stick out his tongue when asked to do so, but will perform the same motion moments later when wetting his lips. A patient suffering from Parkinson's disease may move slowly and awkwardly as a result of the effects of diminished levels of the neurotransmitter dopamine, and yet in a moment of fright rush from the building at the word "Fire!" In all of these instances the same muscles are called upon, but with varying results depending on whether or not a willed voluntary effort is involved or whether the action is stimulated by emotions.

Similar dissociations are observed in what neurologists refer to as the apraxias, complex disorders of voluntary action that are not related to paralysis. Apraxia is defined as an inability to perform learned skilled movements. The first case on record, encountered in 1900, was an imperial councillor initially considered senile. When using his right hand he was unable to imitate simple hand positions (placing his thumb on his little finger) or pantomime how one might use an object such as a comb or a fork. But he had no trouble doing any of these things with his left hand, a clue that his difficulty was not due to senility or lack of understanding. What was involved was a disturbance in the process of voluntary action. Another apraxic patient, when asked to light a candle, attempted to strike the match on the candle while yet another lit the candle correctly but "forgot" to blow out the match. Both puzzling and fascinating was the fact that these errors occurred only when the patients were asked to *voluntarily* perform the requested actions. A patient who could not comply with the request to voluntarily and deliberately demonstrate or mime the use of a toothbrush in front of doctors in a clinic could brush his teeth quite normally when back in his bathroom.

On other occasions the errors involved improper sequencing of everyday actions. When asked to open a tin of soup with a can opener, the apraxic patient might strike the side of the can with the opener or use the wrong end of the opener. One patient performed quite normally when shown a cup, but sat motionless when requested, "Show me how you would use a cup."

Although apraxia is probably the most frequent behavioral disorder associated with brain disease, it is also so subtle that it usually goes unrecognized. Dr. Kenneth Heilman, a behavioral neurologist at the University of Florida School of Medicine in Gainesville, is one of the world's experts on apraxia. And since apraxia provides a window on what happens within the brain during the exercise of will, I visited Heilman in his office in Gainesville in order to learn more.

"Show me how you would use a scissors," Heilman asks. His patient, a sixty-four-year-old man recovering from a recent stroke which weakened his right side, extends the forefinger and middle finger of his right hand and mimics the act of the blades cutting paper. "That is a common error," Heilman says to me. "It's called the body-part-as-object error." Heilman then makes the correct motion, approximating the curl of the thumb and first finger inserted in the handles of the scissors.

"Now show me how you would wave good-bye." The patient looks as if struggling but doesn't move. Heilman then picks up an imaginary hammer and pretends banging a nail. "You do what I do," he says to his patient. The patient mimics instead the act of turning a screwdriver. "No. We did that a few minutes ago." Heilman turns toward me and comments: "When asked to do something different, the patient with apraxia frequently repeats what he just finished doing. We call it 'perseveration.'"

An apraxic patient when encountered for the first time puzzles, amazes, and even amuses, if we can include "black" humor here. For example, when shown a nail partially driven into a piece of wood, the apraxic patient who is offered an array of tools to finish the job may select a screwdriver rather than a hammer. Even though easily able to name a tool when shown one, he may be unable to describe the tool's function.

"Such patients suffer from what we call 'conceptual apraxia,' a defect in our most basic knowledge about the world," says Heilman. With conceptual apraxia, the patient suffers from a defect in the knowledge required to use tools and objects successfully. And at times the disorder can be quite bizarre, as with a patient of Heilman's who attempted to brush his teeth with the toothpaste tube.

Conceptual apraxia strikes one as particularly strange because it's natural to associate an object with its function. Indeed, most of us can't separate those two things even if we try. We see a fork and we think of eating, a chair and we want to sit down. Psychological tests of "free association" depend for their validity on the fact that almost every word in our language brings to mind some kind of functional association (assuming, of course, that we know what the word means). Not all of us come up with the same association, and the more creative person often comes up with some highly unusual ones. But the point is that all of us automatically link our knowledge about the things around us with their use. (It's true that Zen and other transcendence-seeking disciplines suggest ways of uncoupling the knowledge-function dyad. But for most of us, staring at a rock garden for more than a few minutes is more likely to induce boredom than to uncouple our learned association of sand with an afternoon spent at the beach.)

For the patient afflicted with conceptual apraxia, reality is fundamentally altered. It's as if an evil gremlin acting in an instant has returned him to an early state of infancy.

Other apraxic patients demonstrate equally disabling but less global disturbances in their knowledge about what things are and what to do with them. Their difficulty is exemplified by the patient I mentioned a moment ago who used his index and middle fingers to mimic scissors. He also frequently holds the scissors pointed ninety degrees away from the paper he has been asked to cut. When using a screwdriver he also makes errors in the way he uses his body. When pantomiming the use of a screwdriver, for instance, he may rotate his arm at the shoulder joint rather than at the elbow, a movement combination which results in moving the hand in a circle rather than the correct motion of rotating it on its own axis. Heilman has captured these "production errors" by attaching light-emitting electrodes to the joints, recording the movements on film, and then analyzing the picture with the help of computers. This high-tech strategy also turned up oddities in the timing of movements which can't be seen with the naked eye. The patient suffering from this form of apraxia takes overly long getting started and makes a series of brief stops, "stuttering movements," as Heilman refers to them.

Apraxia results from an injury to the left side of the brain. Different forms of the disorder follow, depending on the location of the damage. Neurologists like Ken Heilman are expert in correlating the type of apraxia with the site of brain damage. This involves some of the most complex functions performed by the most complicated organ in the known universe. But however daunting this may at first appear, let me see if I can give you a feeling for how they correlate these strange disorders with the brain.

In 1907 two neurologists encountered a patient who as a result of his stroke could not comply with verbal requests that he pantomime learned skilled acts with his left arm. Since the man's stroke had paralyzed his right arm and not his left, this apraxia involving the left arm was hard to explain. Autopsy revealed damage to the corpus callosum, the band of bridging fibers that connects the two cerebral hemispheres.

Damage to the corpus callosum disconnects the left hemisphere, which comprehends and formulates language, from the right hemisphere. Because of this, the patient with such damage will be unable to comply with spoken requests to carry out movements with his left hand (which is controlled by the right hemisphere, you will recall), yet in the absence of language will be able by imitation to perform correct movements with the left hand.

The neurologists speculated that the left hemisphere contains movement formulas—the spatial and temporal representations of movements. Thus banging with a hammer or writing with a pen takes place by means of the activation of the appropriate movement formula. But their patient was not able to transfer these movement formulas from the left hemisphere to the right hemisphere which controls movements by the left hand, hence his left-sided apraxia.

In 1982 Heilman and two coworkers described a network in the left hemisphere responsible for skilled movements. Starting from the back and working forward, we encounter the inferior parietal lobe in which are stored the movement formulas (see diagram 5). Damage here leaves the patient unable either to follow verbal requests ("Show me how you would pick up and dial a telephone") or to imitate the doctor as he pantomimes the action. If the damage is farther forward, sparing the parietal lobe, the movement formulas exist but cannot be conveyed forward to the motor areas from which all movements take their origin. Think of the representations in the parietal lobe as polydimensional codes that contain all of the spatial and temporal details for skilled movement(down to the level of millimeters and milliseconds). This coded message is carried forward, transcoded, and acted upon. If the brain damage occurs in the motor areas toward the front of the brain, the patient is unable to comply with requests for skilled movements but retains the ability to judge correctly whether or not his doctor, when pantomiming, is doing it correctly. This is an example of a dissociation, a term we will encounter frequently. Since it is so important, here is another example.

In some right handers, like the one just described, movement representations occur in the parietal lobes of both the right and left hemispheres. In other people the "praxic" center, if I can use such a term, is only in the left parietal lobe. In this case brain damage may injure both the language center and the "praxic" center. When asked to pantomime with his left hand, the patient will merely stare uncomprehendingly at the doctor. When imitating or using an object, he makes a variety of apraxic errors: pointing the scissors at right angles to the paper; delays before beginning; the "stuttering" movements mentioned previously.

An area near the front of the brain, the supplementary motor area, plays a particularly important and intriguing role in skilled movements. If this area is stimulated in a patient undergoing brain surgery, he or she will involuntarily make a complex movement of the fingers, arms, and hands. This is thought to be based on the rich connections existing between this area and the parietal neurons which, as mentioned, contain the praxic center. Presumably the representation for the skilled movement is conveyed forward to the premotor area where it is converted into a motor "program" and then forwarded in turn to the motor

Diagram 5

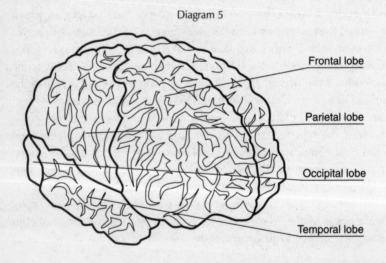

Frontal lobe

Parietal lobe

Occipital lobe

Temporal lobe

area. Here the "program" is run and signals sent downward to the appropriate muscles so that the movement can actually be carried out. You will note that I employ a computer metaphor here (a program) and yet at other points in this book I poke a good deal of fun at the claim that the brain can be understood as nothing more than a megacomputer. This isn't really a contradiction: In some ways and for certain tasks the brain really does seem to function in computerlike ways. Or is it that we have all become so accustomed to computers by now that we just naturally reach for computer analogies? In any case, since I have introduced the supplementary motor area and its role in willed action, let me mention a provocative experiment involving this area carried out in Scandinavia.

Cerebral blood flow to a particular brain area can be measured and the results translated into a color-coded map. Those areas that are most active and therefore use the greatest amount of oxygen are depicted in red, while those brain areas that don't use much oxygen are coded in green or pale blue. (This color distinction is not as arbitrary as it seems; in all of the languages in which I have been able to track the matter, red is a "busier" color than blue or green. If you have doubts, check with your interior decorator.)

When a simple movement like raising a finger is requested of the subject and the cerebral blood flow is measured, the flow increases in the opposite motor cortex. But if a skilled movement is requested ("Show me how a soldier salutes his commanding officer"), increased blood flow occurs in the opposite cortex and in the supplementary motor area on *both* sides. The complex movement involves a "program" which synchronizes the actions of many muscles. Damage in this area results in apraxia, as mentioned a moment ago. But here is the really amazing finding: If the subject is instructed not to make a complex movement but only *think* about doing so, indeed not to move a single muscle, blood flow is increased to the supplementary motor cortex but not to the motor cortex on the two sides from where the actual muscle movements are carried out. This experiment, which has been confirmed by others, sounded a death knell for those few diehards from the Behaviorist camp who deny the reality of the mental. Obviously thinking about making a movement at a time of one's own choosing and altering one's own brain operation in the process cannot be explained on the basis of a stimulus followed by a response.

* * *

Apraxia provides a window on the organization of the brain. Rather than things proceeding apace, with one activity influencing and determining another, brain activity consists of the activation and bringing into play of various sequences arranged in a modular fashion. Some modules are concerned with concentration and will, others with the execution of what one has already decided to do. And as can be observed with patients suffering from apraxia, brain damage can result in counterintuitive and unexpected failures in our mental activities. Intention is disrupted, the desire to do something frustrated by a failure in organizing the movements required to achieve the goal. My point here is that smooth and unified action is made up of separate components and, in the case of brain damage, can break down into these components. But however bizarre the apraxic's performance, his consciousness remains unified: He *knows* something is wrong, even though he can do nothing to correct it. To this extent the unity of the personality remains unaltered: The patient is as amazed and incredulous at his failures of performance as is his doctor. But what happens if awareness is absent? Suppose the patient makes glaring and inexplicable errors of thinking and judgment and yet remains unaware that anything is amiss? To explore this situation and its consequences for the modular theory of brain organization, let us return to where we left off in the last chapter in our discussion of the visual system.

CHAPTER FOUR

An Existential Illness

The poet Seneca wrote to his friend Lucilius: "You know that Harpastes, my wife's fatuous companion, has remained in my home as an inherited burden. . . . This foolish woman has suddenly lost her sight. Incredible as it might appear, what I am going to tell you is true: She does not know that she is blind. Therefore, again and again she asks her guardian to take her elsewhere. She claims that my home is dark."

A case of insanity? A psychotic denial of the painful reality of blindness? This seems unlikely since Harpastes' strange behavior was not accompanied by other accompaniments of insanity or intellectual deterioration. She was still the same person she had been previously, but with the added feature of her blindness and this peculiar and puzzling adaptation to it.

Not much was made of this curious situation until nearly two thousand years later when the neurologist von Monakow encountered a similar denial of blindness in a seventy-year-old man who, like Harpastes, blamed his difficulties on poor lighting. An examination of this man's brain after death revealed destruction in the posterior regions, areas we now recognize as containing the visual cortex. While this explained the blindness, it failed to account for the denial. Persons blind from birth, or at a later point in their lives, rarely deny their affliction. They may become depressed, sometimes profoundly so, but they do not deny their blindness. Something additional is required for denial, but what?

Soon other patients turned up who denied other disabilities. Some denied paralysis of one side of their body, or a partial loss of vision confined to one side of the visual field (almost always the left side for both). Despite the absence of an explanation for this strange propensity for denial of patently obvious disorders, neurologists came up with a name for it: anosognosia, the brain-damaged patient's failure to recognize or appreciate his deficits. Typical of this disorder is the patient encountered by neurologist Edoardo Bisiach.

At age sixty-five, Helen French suddenly lost the use of her left arm and leg. When seen shortly afterward by the doctor at the local emergency room, her mental state seemed perfectly normal, other than for some understandable anxiety. When questioned, she claimed weakness and tingling of her *right* limbs. Even more puzzling than what her doctors initially attributed to right-left confusion was her claim that her left hand did not belong to her, but to another patient who had left it in the ambulance. She readily admitted that her left shoulder belonged to her and therefore, by inference, it seemed likely that the connecting arm and elbow were also hers. In regard to the forearm, she wasn't certain of ownership but insisted that the left hand wasn't hers. At this point Doctor Bisiach placed her left hand on the right side of her trunk. Why were her rings on this stranger's hand? Bisiach asked her. She had no explanation but seemed untroubled by the contradiction. In fact the left arm and leg seemed no longer to exist for her. She ceased doing anything requiring the use of both hands.

Two days later the symptoms began to disappear. She told her doctors that early that morning she recognized the left hand as her own while still recalling her vivid denial of the days before. A CAT scan of her brain revealed the aftermath of a stroke in the right hemisphere in an area containing nerve fibers concerned with feeling and motion of the limbs on the left side.

Note that Dr. Bisiach's patient showed no signs of mental deficiency or instability. She was perfectly alert and not confused or disoriented in any way. Yet she denied something that was so obvious as to hardly need demonstration: the ownership of her own hand. What's more, this incredible assertion coexisted with an openness to reason and persuasion but only *up to a point*. The shoulder and arm—yes, they belonged to her; the elbow and forearm—probably; the hand—definitely not.

Was Helen French aware of her disability? Her initial responses indicate she was not. Yet during the period that she proclaimed this belief she remained in her bed, as would be expected of a person with paralysis of one leg. So in a way it would be fair to state that she both affirmed and denied anything wrong. We have here a split within her consciousness. When asked if anything is wrong with her leg she says "No," but if asked to get out of bed and walk, she refuses or makes an excuse.

When first encountering a patient with anosognosia it is easy to come to the cynical conclusion that he is only pretending and really doesn't believe his own fantastic claims. Such cynicism on the doctor's part can result in tragic consequences for the patient. A patient with brain injury to the visual areas on both sides of the brain told me repeatedly she could see quite normally even though the location of her brain injury made useful vision impossible. Frustrated at my inability to make any headway in convincing her of her blindness, I suggested that she prove she could see by getting out of her bed and walking. At this she promptly leaped from her bed and ran squarely into the nearest wall, breaking her nose in the effort.

Psychiatrists unfamiliar with anosognosia often mistakenly make the diagnosis of a delusional disorder. While it's true that a delusional person may deny any illness along with other aspects of reality, anosognosia is not accompanied by other indicators of mental illness. Other specialists have declared the strange condition a self-protective reflex against the painful reality created by brain damage. Against this interpretation is the fact that anosognosia usually appears only after damage to the right hemisphere and very rarely after strokes damaging the left. Since a stroke on either side of the brain has devastating consequences, why would a self-protective reflex develop only in response to right hemisphere damage? Finally, anosognosia is unheard of with damage to any organ other than the brain, despite the fact that illnesses and injuries in other parts of the body can on occasion have even more destructive consequences. Further, the brain damage responsible for anosognosia involves a breakdown in conscious awareness that cuts across different domains. Helen French not only denied ownership of her left hand, but also acted as if objects off to her left did not exist.

In one test of her left-sided neglect, she was asked to pick up a series of small cubes set out in front of her from left to right. She used her right hand, started from the rightmost cube and continued until arriving at a cube directly in front of her. She then pushed the remainder leftward and exclaimed, "There aren't any more."

Rather than simply a defect in awareness of her left hand and left visual space, Helen French's illness was more existential in nature. It was as if some malevolent fairy had destroyed a portion of her brain and thereby robbed her of the very existence of things and people to the left of her personal space. She was condemmed to live in a reality lacking a leftward component. Thus Helen French doesn't so much ignore or neglect things and people off to her left. Rather, as a result of her stroke they do not exist within her profoundly altered representational system, her inner world.

And there are other convincing reasons for belief in the existential nature of the defect in anosognosia. In most instances, the disappearance of the anosognosia leaves in its wake subtle indicators of dysfunction which the patient could not feign and which can only be elicited by subtle testing. For instance, after Helen French regained her ability to see to the left, she was given what is called the *double simultaneous extinction test*. In this test, the examiner holds up one or more fingers and asks the patient to count them. When the fingers were restricted to Helen's right or left visual fields, she counted them correctly. But when the examiner used both hands and simultaneously flashed fingers to both Helen's right and left visual field, she didn't see the fingers to her left: she "extinguished" them, as the neurologists refer to it. A similar "extinction" occurred if she was touched simultaneously on the back of both of her hands. At such times she only perceived the touch on the right hand, even though she detected touches on either hand if the touches were applied one at a time.

What seems to happen during recovery is a gradual widening of the field of awareness. Starting from a complete loss of function (vision or movement or both) along with the denial of loss, the patient recovers in a series of steps starting with the gradual and guarded admission that—in the case of Helen French—the hand "probably" is one's own, and progressing from there to a defect so subtle that it totally escapes the patient and the doctor too unless he employs subtle tests. Most fascinating are the implications such a disorder creates for our traditional ideas about how we formulate and test our beliefs. We assume that they are somehow interconnected and available for examination and comparison with other beliefs. But in the presence of anosognosia, beliefs can become modular and exist independently and even in contradiction to one another.

The behavior of anosognosic patients shows that the dynamics of beliefs may be viewed differently by the cognitive biologist and the logician. It shows that certain (wholly irrational) beliefs, rather than being inferred from other beliefs, may settle down and subsist decontextualized in a subject's mind; they also suggest that a check for coherence is a secondary process that may or may not take place, even in cases in which the network of all other beliefs is intact. . . .

according to Edoardo Bisiach.

Anosognosic patients remind us that conscious experiences can be dissociated into contrasting and bizarrely contradictory states. In a clear state of consciousness, the afflicted person believes something that is patently incorrect, even ludicrous, and strange to others. Moreover the patient retains this belief despite proofs to the contrary, as when Bisiach held Helen French's hand in front of her. She saw it as a hand, indeed recognized her own rings, but disowned the alien hand. With another patient who suffered a similar left-sided paralysis, Bisiach leaned over from behind him and placed his own hands on the bed, one on either side of the patient's paralyzed alien hand. "Whose hands are these?" the doctor asked. "They are your hands," the patient responded.

"But have you ever seen a three-handed man?"

"No. But," (looking down at the three arms before him) "you obviously have three arms, so you must also have three hands."

In order to explain such a strange disorder, Kenneth Heilman suggests a "feedforward" mechanism in the brain involving a "comparator," or monitor. This hypothetical module receives signals from the intention-motor activation system that initiate arm movement, and detects arm movement by a feedback loop involving signals from the nerves in the arm. The comparator then compares the intended movement with the final position of the arm. Under ordinary conditions there is a match: The comparator receives signals back from the arm indicating it has reached the intended position. If the arm is paralyzed, no movement takes place, no signals of movement return from the arm, and the comparator "tells" the person he is paralyzed. But if those areas of the brain responsible for activating the motor area are destroyed, the comparator never receives signals indicating an intention to move the arm. As a result, the paralyzed person does not know of his paralysis and therefore denies it. As another way of looking at it, updated information of the paralysis is never incorporated into the internal body image. So strong is the influence of the internal body image that it overrides the evidence of the person's own senses!

It's likely that modules like the comparator system are not limited within the brain to monitoring and evaluating just our physical state. They may also influence such mental activities as our decisions, opinions, and attitudes. Even though the comparator operates outside of conscious awareness, we can often infer its influence by the intensity of our inner feelings. When we become "convinced" that an idea or course of action is the "right" one, we often say such things as that it "feels" right to us. At such times we speak as if some inner comparison has taken place, concluding with an inner visceral "feeling" that one conclusion over all others is the right one. William James, although he never spoke of modules or comparators, made a similar point in his essay "The Sentiment of Rationality." How will a philosopher recognize the "rational conception" of reality which he seeks? asks James.

> The only answer can be that he will recognize its rationality as he recognizes everything else, by certain subjective marks with which it affects him. When he gets the marks, he may know that he has got the rationality. What then are the marks? A strong feeling of ease, peace, rest, is one of them. The transition from a state of puzzle and perplexity to rational comprehension is full of lively relief and pleasure.

But the comparator can be wrong, wildly misdirected as with Helen French and others with anosagnosia who deny the physical effects of brain damage. Could similar aberrations take place in the mental realm? Certainly something very similar to anosognosia seems to be involved in the thinking of the paranoic. With the exception of a singular, sometimes deeply hidden delusion, every other belief in the paranoiac's mental repertoire may be perfectly normal. And once the delusion is unearthed, it may remain totally resistant to persuasion, drugs, even electroshock. That's because the delusional belief "feels right," produces within the mind an inner sense of conviction. The comparator is somehow disconnected from the normal feedback which recognizes irrational beliefs at their inception.

A similar process occurs in patients who have undergone surgical disconnection of their left from their right hemisphere to prevent the spread of epileptic seizures from one hemisphere to the other. Michael Gazzaniga refers to the operation of an "interpreter" in the left hemisphere which, like the comparator, operates outside of awareness. He postulates the existence of the interpreter on the basis of experiments like the following:

If the command "Walk!" is projected on a screen so as to be seen only by the right hemisphere of a person with disconnected hemispheres, the subject will get up from his chair and start to leave the room. When asked where he is going, the response is typically something like: "I'm going out to get a Coke." In this instance, since the right hemisphere is surgically prevented from conveying the message "Walk!" to the left hemisphere, it (the left hemisphere) quickly makes up a plausible theory to explain getting up from the chair and walking.

On other occasions, the left hemisphere interpreter provides explanations for emotions. If the command "Laugh!" is quickly presented to the right hemisphere, the subject, after laughing, will not know why he laughed and will say as did one patient, "You guys come up and test us every month. What a way to make a living."

Whatever the mechanism involved, the "interpreter," like the "comparator," exerts a hidden influence on our lives. For instance, our unwillingness to change our minds on issues, our "closed-mindedness" even in the face of evidence to the contrary, is based, Gazzaniga believes, on the operation of "higher inferential mechanisms" like the "interpreter" (and, I would add here, the "comparator").

"We human beings, with our powerful tendency to create and maintain beliefs, readily generate causal explanations of events and actively seek out, recall, and interpret evidence in a manner that sustains our personal beliefs," says Gazzaniga.

Think about the "interpreter" and the "comparator" the next time you are totally convinced that only you are right and everyone else wrong on a controversial question. Oliver Cromwell had an appreciation for the dangers of relying on inner feelings rather than testing out one's presuppositions when he implored the members of Parliament: "I beseech you, Sirs, in the bowels of Christ to consider that you may be wrong!" Our claims to be "logical" and "reasonable" often mask the operation of modular processes that are inaccessible to our awareness, and persuade us not on the basis of facts but by inducing in us seemingly incontrovertible feelings of certitude. Such insights into the operations of our mind should induce humility. They should also deliver us from simplistic and egocentric notions about the "power" of our minds. We are explanation-driven creatures. When an explanation isn't readily forthcoming we put our faith in magic or ritual. Our language, too, is permeated with causal and explanatory terms. In fact rather than language forming the basis for objective disinterested views of things, it too can be traduced and made the slave of mechanisms like the comparator and interpreter. Our greatest challenge in seeking to understand ourselves involves avoiding what Wittgenstein calls the "bewitchment of our intelligence by means of language." But can we depend on language to bail us out of the existential dilemmas created by new discoveries about the modular nature of the brain? To find out, let's explore now how language is organized within the brain.

CHAPTER FIVE

"Is a Bee Smaller Than a House?"

In all spoken languages, meaning is conveyed by means of critical sounds termed *phonemes*. Corresponding to the letters in written language, phonemes are few in number (no more than about 200 distinct sounds across all of the world's languages and only about 38 different phonemes in English). In our own language, meaning is conveyed principally via consonants and vowels (tone plays a lesser role). Vowels are longer in duration, relatively simple in structure, and produced by the action of our nose, mouth, and throat in shaping the vibrations produced by our vocal cords. The tongue, lips, and jaw form the consonants by controlling and shaping briefer, more explosive air bursts. All of this works so smoothly under normal conditions that you and I can distinguish words such as "bat" and "pat" and "cat" even when they are spoken against the noisy background of a crowded room. But those with hearing impairments often fail to distinguish between similar sounds like "d" and "t." To make up for their deficit, they rely on the context in which a word sound appears or they may develop a proficiency in reading lips.

Speech can be heard as pure sound, as when someone speaks to us in a language we do not understand. But even then speech is different from other natural sounds, and most of the time we recognize that we are listening to a language, albeit a language we may not understand. In pure word deafness, fortunately quite rare, speech is appreciated simply as sound. Although the sufferer hears perfectly well, and his understanding of nonspeech sounds such as music or the ringing of the telephone or doorbell remains unimpaired, his native tongue sounds to him like a foreign language. Writing, reading, and speaking are intact, or nearly so. Only the understanding of speech is affected, especially the perception of consonants. "Language sounds like the wind in the trees, a murmuring of a foreign language," as one patient said of his condition. Another said that words to him meant no more than "the rustling of leaves on a tree." Neuropsychologist Kenneth Heilman, whom we met in Chapter Three, told me of a patient with pure word deafness.

One morning while speaking on the telephone with his wife, Samuel Jones (not his real name) suddenly grew angry because his wife had lapsed into "mumbling." When she continued to "mumble" despite his repeated request that she speak clearly, he slammed the phone down. His rage escalated an hour later when his wife, upon her return home, continued to speak unintelligibly. Over the next several hours things deteriorated even further, since other family members called in to restore harmony were also accused of "mumbling." It wasn't until someone resorted to written rather than spoken communication that the difficulty was clarified. The patient (the appropriate designation since he had been brought to a hospital in an agitated state) had lost as the result of a stroke the ability to recognize words. Unfortunately this man not only failed to recover from this grievous affliction, but he lapsed into paranoia. Whereas he had initially interpreted the spoken comments of others as a failure to speak distinctly, to "stop mumbling," he gradually became convinced that others were speaking ill of him or plotting against him. Such paranoia is common among those who lose their hearing late in life. Even though sounds can be heard in "word deafness," they cannot be recognized as speech and therefore, as with a deaf person, comprehension by means of hearing is impossible.

Patients like Samuel Jones suggest the existence within the brain of a specialized language processor, which when it malfunctions results in various forms of aphasia. At its most extreme, things can be even worse then they were for Samuel Jones. A person can suddenly and permanently lose *all* language function as the result of a stroke. Such a catastrophe happens rarely, thank God, but when it does it leaves in its wake a man or a woman, sometimes surprisingly young, with little or no ability to understand, speak, read, or write. But that unfortunate person is not without mental abilities. While deprived of virtually all language capacities, a condition neuropsychologists refer to as global aphasia, he can still think, plan, feel, and express emotion, and communicate by means of facial expressions and gestures. He is not mute, mind you, simply unable to produce prepositional speech. He can and does communicate by mean of grunts, recurrent consonant-vowel syllables ("dada"), and a series of stock words and phrases, including most especially expletives.

A priest under my care many years ago for global aphasia would periodically spew out a string of curse words followed by blushing and other indications of embarrassment. His brain damage included destruction of the input routes for hearing words, the motor output routes for speech, and additional structures around the perisylvian region of the left hemisphere responsible for language. In addition to periodic profanity, he could also sing certain phrases which he could not speak. These preserved language capabilities presumably arose either from older, more "primitive" subcortical regions, or from the undamaged right hemisphere. In either case, his limited speech performance in the face of destruction of the "language center" within the left hemisphere suggests two separate language systems: the dominant, fully operational one in the left hemisphere and a more limited system in the right hemisphere.

Another way to think of it is to consider language as simultaneously conveying two different kinds of information: linguistic, comprising the phonemic and semantic coding by which spoken meaning is conveyed; and the affective or emotional component, conveyed by pitch, tempo, and tone. Persons with damage to their right hemisphere retain their ability to speak and understand the speech of others, but they have great difficulty processing emotions. Some right-brain–damaged patients cannot distinguish sentences spoken in tones of anger, elation, or sadness. Others cannot convey any of these emotions in their own speech, which comes across sounding flat and listless. The problem involves emotional perception and expression, not any difficulty in the *concept* of emotion. If a story is read to them in a flat tone with the reader eliminating any emotional words (sad, happy, angry), they can correctly identify the emotion conveyed by the passage. Thus the person with right brain damage inhabits a world where he can feel and experience the mind's major emotions but he cannot verbally express them or recognize and appropriately respond to emotions in others. A rare combination? I think not. Milder forms of the disorder account, I am convinced, for many of the misunderstandings and grievances that arise involving people with subtle degrees of impairment, especially when they interact with others highly skilled in expressing emotions and deciphering the emotions of others. In describing people with the milder forms of this difficulty, words like "insensitive" and "obtuse" are often employed by spouses, friends, and coworkers, no doubt a response to the flat tones and matter-of-fact manner that accompanies what should be highly emotional topics. The afflicted person meanwhile reacts with an understandable anger, since he *knows* that he experiences emotions but he has great difficulty conveying this fact to others or "reading" the frustration and anger that is being directed to him because of this failure.

Specialists have argued since the time of Paul Broca and Carl Wernicke, the discoverers of the two most frequently encountered forms of aphasia, as to whether the aphasic patient actually "knows" what he wants to say and cannot, or whether he has lost his capacity to think and therefore has nothing to say. Some insight into this question can be gleaned from an examination of nominal aphasia, a form we all suffer from on occasion.

Recall the last time you found yourself embarrassed at your temporary inability to come up with the name of a simple common word like (to take a recent personal example) leash. "Where is the dog's . . . I want to take him for a walk. . . . I can't find the . . . you know . . . the thing to hook on his collar." Temporarily unable to come up with the correct word, most people do as I did and resort to a description of the object or its use. A defect in meaning is not involved here. I knew what object I wanted; I simply couldn't remember its name. Moreover, the elusive word can be speedily retrieved if a clue is provided by another person, perhaps by their uttering the first syllable of the word. And most of the time the word springs to mind a few seconds later with or without outside help. Even if it doesn't, we know perfectly well what object we are after and take up the leash as soon as we see it.

If the word-finding difficulty worsens and becomes habitual and readily obvious to others, it is described as progressive primary aphasia. Most experts believe more is involved here than merely a worsening and deepening of nominal aphasia. For one thing, progressive primary aphasia is not a normal phenomenon, seems to be unrelated to fatigue and stress, both of which make nominal aphasia worse, and progresses to total incapacity. Perhaps the best way to illustrate the extent of the differences is to tell you of a patient of Dr. Andrew Kertesz of the department of neurological sciences at the University of Western Ontario.

At seventy-eight years of age, Martin Tholer, a retired sales manager, complained of a progressive problem with his spoken language that he could trace back at least two years. He continued to manage his own affairs, drive his car, read and enjoy books, and play a respectable game of bridge. Only his language was affected. Here is Kertesz's description of Mr. Tholer. "He looked well for his age and there were no abnormal neurological findings except for a communication disorder. His speech fluency was generally reduced to short sentences with poor grammatical structure or phrases consisting of single words with long pauses in between. He had at times fluent utterances, especially automatic sentences. He had word-finding difficulty . . . but most often he was mute, not uttering any word. His language understanding was much better, which was especially obvious when he responded correctly to multiple-choice questions. He was fully cooperative and easily recalled recent and remote past events."

Notice that despite the extent of his disability, Dr. Kertesz's patient never lost the ability to carry on with his life in every area other than language. His intelligence was not affected (support for the point that thinking can take place in the presence of severe language impairment) and no one could guess anything wrong until Mr. Tholer started to speak.

Patients with Broca's aphasia also seem to "know" what they want to say, as evidenced by their impatience with other people who don't immediately catch on to their telegraphic utterances, along with their evident relief when their listener finally understands what they have been trying so hard to say. The person suffering from Wernicke's aphasia neither understands the speech of others nor produces comprehensible speech himself. But, as with the Broca aphasic, he knows what he wants to say, can even indicate what he wants or intends by means of pictures. Meaning remains intact, as evidenced by the patient's successful managing of all other aspects of his life not involving language. But persons like Samuel Jones who suffer from word deafness lose their capacity to understand what is said to them. For them, meaning is gravely affected, since speech is just another sound in the environment. Nonetheless Samuel Jones can read and, if others write notes instead of speaking, he can understand. But what of persons who cannot read or write or can only do so to a limited extent? Must this compromise their ability to think? Since language is intimately linked with most of the activities we consider uniquely human, it would seem that thought cannot exist without some form of language. But is this true?

At an international conference devoted to "Thought Without Language," a young university mathematician with a severe problem in reading and writing (dyslexia) but with normal speech described his ability to think without language. "From an early age I found that many things are easier to think about without language." When working out the total resistance of a network of resistors, he imagined the physical pattern and ". . . manipulated the resistor network by mentally cutting, folding, and reconnecting it. . . . The process was completely nonverbal, though just as precise as the algebra for which it was a substitute. From an early age I found that many things are easier to think about without language."

Although the young man's experience suggests that thought can exist without language, intriguing alterations in both processes occur secondary to brain damage. In an experiment carried out by Michael Gazzaniga, a man who had lost sight off to his left because of brain damage to the visual area on the opposite side was asked to imagine himself looking toward California from New York and naming the states in between. He named only ten states, all located to the right of his imaginary vantage point. He omitted the states to the left, corresponding to the visual field mediated by his right brain lesion. What he could not see in the real world as a result of his brain damage, he could neither picture in imaginal space nor speak about. Such selective visual processing failures may even result in the incorrect reading of words. Thus patients who neglect everything off to their left as the result of right-sided brain damage may also misspell the left half of words or, when writing to dictation, begin from the last letter and proceed backward from right to left, often omitting some of the leftmost letters. One such patient, after writing "BUE" for BLUE, spelled the word correctly when asked to write the word in mirror-reversal. The beginning of the word could only be represented if it could be mentally projected into the normally perceived right half of space. Some people with these difficulties have described their inner experience. One said attempting to spell a word was like looking at an image in one's mind in which the letters on the right side appear clearer than those on the left. Another with a similar defect said he could tell how many letters were missing at the beginning of the words he had written. Similar difficulties may affect speech as well. The patient who wrote "BUE" for BLUE would say "balance" for "ambulance" or "portent" for "important."

Such performances suggest to Edoardo Bisiach, a neurologist in Milan, Italy and the specialist in neuropsychology who cared for Helen French whom we met in Chapter Four, the ". . . radical hypothesis" of an implicit dependence of language on visuo-spatial organization within the brain. Damage to parts of the brain serving vision and space affect language as well. This can involve deficiencies or loss, as with the examples mentioned a moment ago, or *productive* forms of the disorder. For instance, the patient may employ substitutions or additions at the start of words. "Train" is read as "brain" or "right" as "bright." In making these substitutions, the patient acts as if he is somehow dimly aware of something wrong off to the left, and reflexively attempts to correct it.

Brain damage and the resulting loss of function can on occasion evoke the elaboration of strange and preposterous claims. This process is called *confabulation*. An example is Bisiach's patient Helen French, who claimed her paralyzed hand did not belong to her but to another patient who had been in the ambulance earlier. This denial of illness, anosognosia, may also take more attenuated forms such as denial of loss combined with a tacit awareness (ignoring and not using the paralyzed side, as with Helen French). But what is most intriguing about anosognosia, no matter the degree of its elaboration, is its encapsulated nature. Everything about the patient's mental life seems normal, with this one glaring and discordant area of processing. And—here is my main point—language is no help in correcting the distortion. The person cannot be "reasoned with," and continues to maintain his outlandish claims that nothing is wrong despite direct and seemingly undeniable demonstrations to the contrary. Such resistance to the power of "reason" flies in the face of our preconceptions about the preeminence of language as our most direct personal expression, the measure of all of the other mental processes going on within our brains. But actually language is only one mental function among many. And it can fail to overcome distortions created by brain damage affecting sensation and movement. Moreover, language is also only one module among many. It does not enjoy privileged status but can break down or evidence bizarre distortions.

I described on page 51 a patient's bizarre remark that the doctor must have three hands, in response to the doctor putting his own hands on either side of the patient's paralyzed but disowned hand. At another point the doctor, Edoardo Bisiach, defied his patient to move his left hand as proof that it was not paralyzed.

Patient: *(Hesitates)* Just give me time to proceed from thought to action.

Doctor: Why don't you need any time to proceed from thought to action when you move your right hand? Maybe you *can't* move your left hand?

Patient: I can move it all right. Only . . . sometimes there are illogical reactions in behavior; some positive and some negative.

When further challenged by Bisiach to move his paralyzed hand, the patient hesitated and said: "You see, doctor, the fact that my hand didn't move might mean that I don't want to raise it. My words might astonish you, but there are bizarre phenomena. My not moving my hand might be due to the fact that if I keep from performing this movement I might be in a position to make movements which would otherwise be impossible. I am well aware of the fact that this seems illogical and uncanny. Indeed, this obscurity is repugnant to my mind, which is very rational. I hope I am not boring you, doctor, with my apparently odd talk."

This patient is describing the loss of the ability to form an internal representation of his illness. As a result, he speaks in tentative terms, as if guessing what might be the difficulty. He recognizes on some level that his explanations are "illogical and uncanny," but he cannot break through them. The best he can do is to fight against an obscure repugnant inner perception of himself. Although the patient and Dr. Bisiach can speak to each other, they inhabit different worlds, different realities. The doctor's and the patient's interpretations of the patient's difficulties are completely remote from one another.

Language is only one of a group of selective knowledge impairments. Others include loss of the ability to recognize people on the basis of their facial characteristics. In this condition, known as prosopagnosia and discussed in more detail in Chapter Seven, recognition is based on hearing the person's voice or watching him or her walk.

Thus, recognition of a particular person or thing may be based on any number of characteristics: how she stands or walks, the perfume she customarily wears, the sound of her voice, the feeling evoked by her handshake or kiss. Isolated breakdowns in recognition can occur along any of these lines. Further, the pattern of these breakdowns provide clues about the ways knowledge is encoded within the brain.

Philosophers have speculated for centuries about how knowledge is organized. Most have thought of it as a unity. Thus, if you had knowledge about, say, a violin, then you could rely on sight or sound or touch or other senses (no one so far has described the *taste* of a violin) in order to recognize it. The nineteenth-century neurologist Lissauer described the process this way:

> With the violin's image there are connected a number of recollections which concern its name, its sound, its image. The sound of the instrument, the sensation and tactile experience which go with its handling. In addition there may be the optical image of the violinist in his characteristic pose. It is only when the associations between the percept of the instrument and the above-mentioned recollections occur promptly in consciousness that one is enabled to interpret the object as a musical instrument and differentiate it from other instruments and generally to categorize it. If, however, this association is delayed or disrupted through some pathological process then even if the image of the violin is perceived, however precisely, there are no associations with prior experiences and recognition is therefore not possible.

Lissauer's point was that our knowledge of a violin involves the linkage of various sensations. A disturbance in any one of them can interfere with this process. But in real life this would not happen, because one sensory channel makes up for deficits in another. The deaf recognize the violin by its size and shape; the blind rely on its sound. And only in the artificial testing situation of the neuropsychologist's laboratory would a deficit in one sensory channel interfere with this process.

Implicit in Lissauer's example is the unitary concept, "violin," contributed by all of the various sensory channels. Thus hearing the spoken word "violin," or looking at the instrument, or even just picturing it—all are encoded in the same neural network within the brain.

According to this theory, all sensory channels "feed into" an area responsible for the unitary concept, "violin." Once the concept is accessed through one sensory channel, it becomes available to all others. If I recognize a violin by its sound, then according to this theory it would be impossible for me not to also recognize the instrument from seeing it, or hearing the word "violin."

In actuality, the organization of knowledge within the brain does not correspond to this intuitively appealing theory.

Neuropsychologist Alfonso Caramazza of the Johns Hopkins

University showed in the mid-1970s that failures in understanding sentences can result from either damage to lexical processes (the understanding of words) or from damage to syntactical processes (the way words are organized in a sentence). Our knowledge of syntax, for instance, enables us to distinguish between "The cat was chased by the dog." and "The dog was chased by the cat." The difference in meaning in these two sentences is conveyed by the grammatical relations existing among the words. By meticulous and sensitive examinations of patients with brain damage, Caramazza and coworker Edgar Zurif found that syntactic processes may be disrupted while knowledge about the words making up the sentence remains unimpaired. For example, a patient after a stroke will continue to understand the meaning of "cat," "chased," and "dog," but will fail to understand the pursued-pursuer relations in the two sentences.

This dissociation between word meaning and word placement within the sentence suggests, says Caramazza, that these two kinds of linguistic knowledge depend on distinct structures or processes within the brain. "Further, by studying brain damage and the resulting impairments in performance we can reconstruct how the whole system works. This is in fact the challenge: how to understand the structure and functioning of the normal brain on the basis of looking at brain damage?" Behind this search lies an unprovable but logically compelling assumption: When damaged, the brain's performance impairment isn't arbitrary but consists of specific deformations of its normal operating principles.

Anatomy provided one of the earliest operating principles to be uncovered by Caramazza and others. Normal syntactic processing is dependent on neural structures in the frontal-temporal parts of the left hemisphere, while the understanding of word meaning seems to be represented in the temporal-parietal regions of the same hemisphere. But even this distinction is "too simple," according to Caramazzo who seeks more precise knowledge about how language is organized in the brain. He has concentrated over the past decade of research at Johns Hopkins on a relatively narrow aspect of language: the use of *single words* in naming, reading aloud, and writing.

"We need to clarify the distinction between knowledge of the meaning of a word and our knowledge of the form of a word," Caramazza suggested to me as we sat in his office on the campus of Johns Hopkins University. "For instance, take the word 'cup.' You know that the word *cup*—the sound 'k ǝp and the letter sequence c/u/p—refers to an object used for drinking. Anyone who speaks and understands English must manage the completely arbitrary nature between the form of the word and the concept it refers to."

Three distinct forms of knowledge are involved, Caramazza pointed out: phonology (sound), orthography (spelling), and meaning (semantics). Normally, if we can access one of these forms we can access the others. Despite this seeming unity, all three types of knowledge of words are represented independently in the brain and, what's more, in the event of brain damage can break down independently. Some hint of this division arises even without brain damage: on occasion we can speak a word, recognize it when spoken by others, know the word's meaning, and yet not be able to spell it.

A common error frequently resulting from brain damage involves producing a semantically related word instead of the correct response. When shown a picture of a dog, the patient may identify it as a cat. The same semantic error may also accompany reading the word "dog" aloud or writing it to dictation. Not all patients make the same kinds of errors. One patient of Caramazza's—known by the code letters K.E.—responded to hearing the word "shelf" by writing "lamp." He offered the same response when reading the word. When shown a picture of a shelf, he verbally identified it as a "chair." When requested to write what he saw in the picture, he wrote "light." In short, he was completely incapable of processing the word either verbally or in writing.

"Such an across the board failure indicates sustained damage to the semantic (meaning) component of the word processing system. Both speaking and writing were affected."

Another of Caramazza's patient's, R.G.B., made errors only when reading a word. When asked to read aloud the word "volcano," for example, he responded with "lava," but when asked to define the word he replied "Fire come out of it . . . a big thing . . . a mountain." He obviously understood the word, could define it satisfactorily when asked about it, but could not recognize the word when he read it. Other patients may show just the opposite response: they could read the word, adequately define it, but not be able to write it.

The performance of patients like K.E. and R.G.B. suggests that the knowledge of the meaning of words is represented in the brain independently from the sound of the word and its written form. This is, of course, at variance with our everyday experience. If we know the meaning of a word we can recognize it on sight, speak it aloud, recognize it when others speak it, and in most instances spell the word to at least a close approximation. But Caramazza's work suggests the existence within the brain of autonomous processing components which can break down in selective and counterintuitive ways. We encounter here yet another example of the fundamental principle Caramazza espouses: " . . . the modular organization of cognitive processes and the brain. Cognitive abilities such as language are the result of the concerted activity of many simple processing mechanisms distributed in many different regions of the brain."

Even finer grained and more remarkable dissociations exist. Neuropsychologists Rosaleen McCarthy and Elizabeth Warrington encountered a patient with a knowledge impairment involving the verbal definition of animals. He did quite well with inanimate things (he could define a lighthouse quite satisfactorily), but when asked to define an animal such as a pig he could only respond "an animal." Left at this, his problem would but be described as a "category specific" defect. But the patient's performance was abnormal in another respect: While he could not come up with a satisfactory definition of an animal, his definition when shown the animal's picture was often elaborate and detailed. Thus when asked to define a rhinoceros, he said, "animal, can't give you any functions," but the picture of a rhinoceros elicited, "Enormous, weighs over one ton, lives in Africa." This discrepancy in performance between word definitions and picture descriptions argues against the idea of a single semantic system shared by all of the senses. Instead, it seems likely, based on patients such as this, that the processing of information by the different senses involves different brain networks and meaning systems. What might be the advantage of such an arrangement?

Imagine driving a car while discussing politics with a companion. Imagine now that your companion begins reading a road map and says to you, "Turn right at the fork ahead." In most instances, such a command should pose no problem. Separate processing of what you see and what she says makes it possible for you to wend your way through difficult driving situations while holding an animated conversation. Your visual system is engaged in one area (driving) while the verbal system is accessing others (economics, politics). But if vision and verbal understanding shared a common channel, the interference effect between what you see on the road ahead and your hearing of what she has just said would increase (as would your chances of an accident!).

Competition between sensory channels can also prove disruptive. Your smooth and effortless driving performance will deteriorate if at some point you begin reciting out loud, say, the makes, years, and models of the cars you encounter. In this instance the automated response to the cars (avoiding them, practicing sensible and courteous driving habits, etc.) will be compromised by the verbal system which is now focused on the same objects, albeit for the different purpose of identifying them. The verbal system may, for instance, need more time to carry out an identification of a particular car and therefore order the necessary muscular adjustment on gas pedal and brake to keep that car in view for a potentially dangerous prolonged time.

It's likely that knowledge within the brain is stored not as a unity (a tiger) but according to separate components or modules (the sight of the tiger, its roar, its smell, etc.). Further, some of these modular components may malfunction without affecting any of the others. Thus I may be able to respond to questions about lions based on general knowledge (Is a lion dangerous?) but not questions that would require visual knowledge (Does a lion have four legs?). In short, "lion" doesn't exist in my brain as a unity but as a multiplicity of such different knowledge categories as vision, hearing, touch, and general knowledge (A lion is a member of the feline category of animals.). One area may be gone while other areas may not be affected. Indeed, the concept "lion" may disappear as part of the loss of every other creature in the category "animal." Or the process may take the opposite form, as with a patient described by speech therapist Jennie Powell and psychologist Jules Davidoff.

N.B., a twenty-seven-year-old woman blinded in an auto accident, could not see an orange "in my mind," but could identify it as round because she retained the ability to remember how it felt. But this inability to internally visualize could not be attributed to her blindness alone: Most persons with acquired blindness retain some powers of visual imagery. And there was another difference from blind people: Her general knowledge about inanimate objects lagged far behind her knowledge about animals. Rather than just a loss of vision, she suffered the loss of whole categories of knowledge. Powell and Davidoff feel that N.B. ". . . has an island of preserved knowledge. It concerns not all general knowledge but only that concerned with animate objects."

In the normal brain (not the best of terms—the "undamaged brain" says it better, but seems awkward), one sensory channel may dominate others: the person who learns better from listening than from reading the same material. Our language captures these preferences: "I see what you mean" versus "I hear what you're saying." And persons reliant on seeing or hearing may come into conflict with those whose reality is more attuned to touch. "She rubs me the wrong way." All of these phrases hint at the uniqueness of each person's subjective world. Teachers know this intuitively. Some students learn little from the usual classroom process of sitting and listening but do well reading the same material on their own. Indeed misfits between a student's preferred sensory channel for learning and an unsympathetic teaching style may lead to learning disabilities, hyperactivity, and conduct disorders like the so-called attention deficit disorder. Here the child—or adult, since it's recognized now the disorder carries over into adulthood—experiences difficulties in concentration, sitting still, attending, and focusing on the material at hand. In dyslexia, the most common learning disorder, the difficulty stems from an atypical organization within the brain of those areas concerned with reading. In compensation for this difficulty, the dyslexic employs alternate means of learning, such as tape-recording lectures for later playback.

Such individual differences suggest to neurologist Anthony Damasio that "A given individual or object generates a multiplicity of representations *within* the sensory cortices of the same modality (for vision, examples are shape, color, texture, motion) and *across* cortices of other sensory modalities (hearing, touch, smell and so on)."

One of Damasio's patients with damage to the ventral occipital and temporal association cortices can readily identify man-made tools when shown pictures of them but fails to identify natural objects. "The same attentive and intelligent patient who quickly recognizes a 'saw,' a 'screwdriver,' a 'shovel,' or an 'electric shaver' agonizes over the recognition of most animals and food items, venturing hesitatingly that perhaps they are some kind of 'animal' or 'plant'," says Damasio. Another of his patients showed just the opposite pattern: trees, flowers, and other objects posed no recognition problem, but he couldn't distinguish the picture of a squirrel from a cat, thinking them both cats because of the whiskers.

Other recognition categories include body parts, indoor versus outdoor objects, proper names for objects. And several patients are on record who could no longer recognize musical instruments from their pictures. They did fine when shown other man-made objects; only musical instruments eluded them (but hearing the instruments triggered immediate recognition). A strange combination of circumstances. Certainly there is no reason to anticipate that knowledge would be organized within the brain according to such categories. Indeed, many of these categories seem outright bizarre: a farmer suddenly lost the ability to recognize several of his animals. Another patient could neither name animals by their pictures nor identify them when she heard the sounds they naturally produce (a dog's bark). She did no better when asked to write her responses, indicating her errors did not stem from a speech production deficit. And while she readily answered that a stool has three or four legs, she couldn't tell the number of legs of a given animal. Yet her perceptual knowledge of animals (whether a sheep is white or green) remained intact. Her deficit involved a disturbance in the relation of language and visual information, a difficulty putting into words the observable characteristics of animals. Even more intriguing are the findings of Barry Gordon, a neuropsychologist at Johns Hopkins University School of Medicine.

A thirty-nine-year-old woman undergoing surgical treatment for her epilepsy agreed to allow Gordon and his colleagues to stimulate her brain with a low-voltage electric current. Before and during the application of the current, Gordon carried out tests measuring her understanding of speech, recognition of objects from their pictures, the reading of words and paragraphs, and her ability to name objects on the basis of verbal descriptions. Trials with and without electrical interference were randomly mixed so that neither the patient nor Gordon knew when the electrical current was being applied. The electrical interference, lasting between 0.5 and 3 seconds, began just prior to a specific test, and ended just before her response. Extensive testing revealed only one impairment. She could not make verbal judgments about size. When asked "Is a bee smaller than a house?" she answered correctly every time when no current was applied, but with the current, her responses were no better than random guessing. But when shown a picture of two objects and asked "Which one is bigger in real life?" she answered correctly one hundred percent of the time both with and without the electrical interference.

"This case extends the classes of categorical cognitive distinctions represented within the brain," Gordon told me during a discussion in November 1992. "The category *size*, discovered to be impaired in this patient is an inferred physical property, not at all like those previously described which have been limited to perceptual categories like color or abstract semantic categories like a fruit, a vegetable or an animal. What's more, this category-specific impairment, affecting only verbal access to visual size information, indicates that the categories of human information can be very restricted and not predictable from prior scientific knowledge."

Gordon and other neuropsychologists are discovering that the brain's organizational makeup does not at all follow "common sense" or intuitive notions. Nor does it divide the world up into categorical distinctions that most of us employ. "The brain divvies things up and puts them in bins. And these bins may be different from what you would expect. The brain is not necessarily built the way your mind thinks it is," according to Gordon. While living versus nonliving distinctions are appealing and seemingly sensible, what use is distinction between musical instruments and other man-made objects? Certainly neither logic nor intuition helps very much in arriving at such a division. Most probably we have evolution to thank for some of this.

"These category-specific deficits seem to obey what might be considered natural categories (such as the difference between animate and inanimate objects), so there might be some basis for believing that evolution has seen fit to dedicate specific neural systems to their representation, which is why they can be selectively dissociated," says Barry Gordon. But he acknowledges that evolution is only part of the story. "Certainly something like a category-naming deficit for indoor objects seems to have little to do with evolution. While it appears that cerebral injury can erase very selective portions of semantic memory, how this can be explained neurobiologically is as yet unclear."

But one can hardly overestimate the importance of learning more about the categories and how they are organized within the brain. They form the underpinning to our understanding of ourselves and the world around us. The contemporary philosopher Oliver A. Johnson, writing about Immanuel Kant, who anticipated many of these findings two centuries ago, says:

> What is the source of our knowledge of the formal elements that give our experience its structure? According to Kant, this structure is the result of

the creative activity of the mind. When the mind receives the input of the senses—the shapes, colors, sounds, and so on—it simultaneously does its work of organizing these raw data into coherent structures or objects in a single unified whole. The results of its activity (of which we are conscious) and that of the senses working together is the coherent world we experience.

Of course, we are as a rule *unconscious*, or at best partially conscious, of the mind's (we would say the brain's) activity. Moreover, I wonder what if anything would be gained if such processes gained entry to consciousness. Awareness is of no help, indeed may be a liability, in many of our performances (walking, driving, choosing each of our words separately rather than speaking spontaneously). Similar situations arise in the psychological sphere. Awareness of and preoccupation with our own and other people's motives frequently leads to misunderstandings, resentments, and other breakdowns in communication. Besides, as our legacy from Freud, we have reason to doubt that full awareness of our motives, drives, and other mental activities may be possible. We are built, it seems, in such a way as to become aware only of *dysfunctions* rather than functions.

And there is another consideration here, a humbling one if we really think about it. The study of disorders may be misleading when it comes to understanding normal brain function. Removal of a part from the engine of a car may interfere with the engine starting, but we hardly conclude from this that smooth automotive performance flows from this part alone. So we must be cautious in assigning to a part of the brain responsibility for the carrying out of a specific function. Nor should we assume that these categories form the basis for our mental life. This is especially true when it comes to interpreting the importance of the categories discovered by Gordon and other researchers. Just because mental activity can be fractionated into specific categories (fruits, inanimate objects, size) doesn't imply that our thinking is necessarily constrained by them. Creativity involves imaginatively blending and even transcending the usual categories. Fairy tales, science fiction, and the works of artists like Magritte are examples. In these productions, plants may talk, objects express emotions, and landscapes harbor humanlike resentments. So we can say with confidence that we are not completely dependent on categories, or at least the categories can be imaginatively and creatively transformed. But the loss of certain categories of knowledge brought about by brain damage does raise important questions about how meaning is embedded in the brain. While written and spoken language employing the same words convey the same meaning (ignoring such anomalies as illiteracy), neuroscientists have long wondered whether written and spoken meaning are processed similarly in the brain. Recent work, also done at Johns Hopkins, provides an answer.

A seventy-seven-year-old woman suffered a stroke affecting her left hemisphere and the anterior portion of the underlying basal ganglia. As a result, she began to ignore objects off to her right. Asked to draw a clock, she left out the whole right side of the clock face. When she attempted to copy drawings she omitted everything on the right side. The reading of words followed a similar pattern, with the omission or distortion of the end of the word. "HOUND" was read as "house" and "STRIPE" as "strip."

At first such deficits might suggest an impairment in seeing the right sides of objects. But she had no problem recognizing and reading all the separate letters of a printed word like H, U, M, I, D, but only with the letters configured separately. When asked to read the letters as a word, she substituted or omitted letters on the right portion of the word, in this instance saying "human." Perhaps her problem involved the *reading* of words? Not really, since hearing the word produced the same result, as did spelling words out loud, or writing them out on paper. She said "Clou" when shown C-L-O-U-D and offered "plane" for the presented word P-L-A-N-E-T. When asked to spell a word backwards, she left out or had difficulty with the first few letters. I-L-L-N-E-S-S was spelled "enlli," omitting "ss," the extreme right portion of the backwards version of the word.

Brenda Rapp and Alfonso Caramazza, the two Johns Hopkins researchers who carried out the testing on this patient, believe their results suggest that the brain forms an internal representation of word spellings (and visual shapes) that exists in mental space and is set out from left to right. The representation is the same no matter whether the word is read, listened to, or spoken out loud. But it does matter whether the word is processed as a *word* instead of just a string of letters (recall that their patient had no trouble with the separate letters H, U, M, I, D, yet when putting them together she read "human"). Damage to the left hemisphere of the brain of their patient resulted in problems in processing the right portion of these internal representations.

Rapp's and Caramazzo's work provides additional support for our contention that the brain is functionally organized to deal with meaning. Specific breakdowns in the processing of meaning occur as the result of brain damage in various areas. As mentioned before, the patient may no longer be able to name members of a specific category such as fruits, animals, vegetables, or proper names. Patients have also been encountered who have difficulties handling grammatical categories, such as nouns and verbs. Some can name nouns but not verbs, while in others the pattern is reversed. Two patients with brain damage studied by Caramazza and a colleague illustrate how the brain might be normally organized for the processing of grammar.

Both H.W. and S.J.D. have difficulty with homonyms: words that look and sound alike but can have different meanings. An example is the word "watch" in the two sentences, "My watch is slow" and "I watch a lot of TV." Both patients did well in reading the first sentence where "watch" is a noun but could not read the same word when used as a verb as in the second sentence, H.W. had his difficulty only when reading the sentence, whereas S.J.D. only had trouble with the verb form when asked to write the sentence. "Such findings suggest that grammatical classes are represented independently at some level of processing," according to Caramazza. Moreover, these difficulties may be highly specific and may occur when the person reads or writes or listens.

Here is a real insight into the relationship of mind to brain. A whole category of discourse, in this instance verbs, can be missing or impaired. Since grammatical structure is the key to meaning, deficits in handling grammar impair a person's ability to extract the essence of a situation. We touch on something very fundamental here. Sages and philosophers have argued for centuries about, to put it somewhat clumsily, the meaning of meaning. What is the relationship of the mind to the categories of human experience? Why do we separate things in the world into living-nonliving, vegetable-mineral, and so on? Based on research like that done at Johns Hopkins, it appears that such categorizations are based on our brain's organization. If our brain were organized differently, it's likely we would inhabit a different world not only of perception and action but also of meaning. And with brain damage, meaning can break down in a host of different ways. Lest this sound too theoretical, consider the patient of J. L. Nespoulos identified as Mr. Clermont.

When speaking or reading, Mr. Clermont left out words and substituted incorrect and inappropriate words. Yet his difficulty remained confined to the production of sentences. Given the words in isolation and one at time, he did fine, but when the words were combined into a sentence, he made errors. He could not distinguish whether a word was to serve as a verb or noun or other grammatical function word. In one test, the examiners tried a trick. They composed a sentence and instead of writing it horizontally across the paper, they wrote it in a vertical column. He promptly read the first few words of the sentence correctly and then exclaimed, "Oh, it's a sentence!" and regressed once again into committing many errors.

One way of describing Mr. Claremont's difficulty is that he had lost the ability to extract meaning from a sentence despite a retained ability to understand the meaning of each of the words in isolation. But this was a grievous loss, since meaning involves not just individual words but their arrangement in a sentence, the syntax. Change the word order or grammatical function of a word in a sentence and the meaning is changed: attend a "fire sale," "fire" an employee, "fire a rifle." The ability to read the word "fire" in isolation isn't very useful in extracting the meaning of these phrases.

Meaning is also heavily dependent on use of function words. Consider these two sentences: "Mary washes cars." and "Mary gave up the job in a car wash where she washes cars." The first sentence contains three concrete words, each of which can be pictured in the mind. The second sentence, along with being more informative, contains several words that relate to the grammatical form of the sentence. These words ("the," "in," "a," "where") are abstract: They do not refer to anything that can be pictured. Speech therapists Sue Franklin and David Howard tell of a patient, Derek, who lost the ability to hear abstract words.

At age fifty-four, Derek suffered a stroke which initially left him unable to speak. One month later he recovered to the point that his speech was almost normal again. But while people could understand him, Derek experienced the speech of others as "mumbling" (you will recall Kenneth Heilman's patient underwent a similar experience). Tests administered by Franklin and Howard revealed that Derek was not "word deaf," that is, he had no difficulty in deciding whether two words or syllables were the same or different. Nor did he have difficulty understanding spoken words as long as they stood for things that could be pictured. His problem involved a difficulty in understanding abstract words. He had no problem giving short definitions of concrete words, but he missed almost half the time when asked to define abstract words ("meek," "idea"). But here is the crucial point: Derek's reading, comprehension, and definition of abstract words was perfect; it was just *hearing* them that he couldn't do.

Imagine for a moment Derek's world. Since he can only understand concrete words, he can only talk about things that can be pictured in the mind. Take the sentences mentioned above. While he will have no problem with understanding Mary washing cars, since all three words refer to concrete entities, he will have no idea in the second sentence that Mary no longer works at the car wash and therefore is no longer washing cars. Yet if he reads, rather than hears, either of the two sentences, his comprehension is perfect. Because of this difference between his auditory and reading comprehension, Derek's conversation lacks variety, subtlety, and zest. More importantly, in the verbal sphere abstractions do not even exist for him. It's not that Derek is ignorant of concepts like "inflation" and "ennui"; it's just that knowledge of these concepts only exists and can only be experienced and expressed by reading about them.

Derek and others like him suggest that spoken and written comprehension occurs in separate areas within the brain. This is of course not at all the way we think of comprehension. When it comes to comprehension of a foreign language, for instance, we would not expect a fluent errorless reading knowledge to coexist with a total inability to speak or understand the spoken form of the language. We encounter here once again the basic modular organization of the brain, in this instance two modules. In the normal brain, knowledge is organized to include both modules operating simultaneously. In Derek's case, one of the modules has been rendered nonfunctional, with knowledge of a particular category (concrete versus abstract) entirely dependent upon the other module. (In Chapter Ten on "split brain" patients, we will say more about the multiplicity that can exist with the brain-mind of a single person.) But I believe Derek and others with similar afflictions partake of the same multiplicity. When reading, Derek is perfectly capable of understanding abstract concepts and can respond in writing employing equally abstract concepts. Yet talk to him about the same abstract material and he hears your speech as nothing more than "mumbling."

So do abstract concepts exist for him? The only real answer is that in a way they do and in a way they don't. The situation is like the double-slit experiment in quantum physics, where the design of the experiment, the questions the experimenter asks, determines the answers that he obtains. If the experiment is set up one way, light is measured as a particle; set it up another way, light is a wave. "Inflation" exists for Derek when he reads it or writes about it but not when it is spoken about. The term does not even exist for him in the auditory sphere. Thus knowledge is not unified or "one" as we have been taught to believe, but consists of modules, any one of which may fail and affect the structure of knowledge in that particular brain.

Derek illustrates not only the modular nature of knowledge but also that of identity and personal integration. Although we experience ourselves as a unity, our sense of oneness depends upon the smooth interaction of several modular functions. Damage to any one of these modules fragments our personal integration in specific ways.

Memory seems to offer some anchoring point here. At the most basic level, we know who we are and maintain our sense of personal identity and integration on the basis of our ability to remember our past experiences. None of us remembers everything, of course—indeed, who would want to?—but most of us remember enough from events over the years to provide ourselves with a sense of our own unity as a single, reasonably well-integrated person. Memory thus forms the underpinning for our personal sense of identity. But brain studies carried out over the past few decades cast doubt on this cheeky confidence in memory as the substrate for our own personal integration.

CHAPTER SIX

Time Stops for Mr. M

L oss of memory is one of the first signs of the dreaded Alzheimer's disease. Eventually the memory loss worsens to the point that the victim seems like a stranger to himself and others, only a ghost of his former self. Accompanying this loss of memory comes a corresponding loss of personal identity. Friends and acquaintances aren't recognized, past events disappear from the mental landscape, the commonest of words prove maddeningly elusive. With further progression, the sufferer slips into a psychic black hole until, eventually, personality impoverishment makes even normal conversation impossible.

Alzheimer's serves as a grim reminder that memory and identity are interlinked. Indeed, what are we but the sum total of our experiences? When we cannot access these experiences we are deprived of parts of ourselves. Like our identity, memories are not static but constantly changing. Yesterday's occurrences can be recalled in far more detail than what happened a week ago, while the events that transpired on the same date last year may be entirely forgotten. Yet at one time last week's and last year's event's were as retrievable as yesterday's. In the interval those memories have undergone modification, amalgamation, reconstruction, or even total disappearance. It is as if a chapter or a paragraph has been altered or is missing entirely from our autobiographical memory. Things we have forgotten may still influence us (we have Freud to thank for this insight) but unrecoverable memories deprive us of self-knowledge. (Ironically, people who are closest to us may recall highly revealing events about us that we have forgotten and thus possess greater insight into our identity and our character than we possess ourselves.)

A carelessly conducted operation on the brain can bring on a devastating loss of memory and, in the process, destroy the patient's identity.

The most famous case was that of a young man, H.M., who underwent surgery on 1 September 1953 in an attempt to improve severely disabling epileptic seizures. After the operation, he could no longer recognize the hospital staff. If his doctors went out of his sight for even a few minutes and then returned, H.M. couldn't recall having ever seen them before and had to be reintroduced. He could still recall those things he knew before the operation, but he could not retain any new information for more than a few seconds. Now in his sixties, he remains unable to form new memories of everyday occurrences, such as what he ate for lunch or the identity of the person who called on the telephone just moments before. In short, he cannot store new memories—those since the time of his surgically induced injury—although he still retains memories of events that occurred prior to the operation, such as his date of birth or where he went to high school. This condition is called *anterograde amnesia* and is limited to difficulties in recalling or recognizing events that have occurred *after* the onset of the amnesic condition. As a result of this amnesia, H.M. lives in a world of fragmented and disconnected experiences. In contrast to H.M., persons afflicted with *retrograde* amnesia cannot recall events that happened *before* the onset of their amnesic condition. Both forms of amnesia disrupt identity in different ways. People with retrograde amnesia can form new memories but feel dislocated from the person they were prior to the amnesia.

For H.M., time stopped on 1 September 1953. Of course he knows it is not 1953—his intelligence is normal, and he retains the ability to interact and converse with others as long as he is not required to store new information for later retrieval.

Testing of H.M. reveals some surprising findings. After reading a list of commonly used words, he forgets them in a matter of seconds. Yet if given the same list at a later time he learns the list more quickly, although once again he promptly forgets it. If shown a list of words, he tends to use these words in later tests for word and sentence completion, but doesn't remember having seen the words before. It's as if he remembers on one level of the mind but forgets at the level of consciously recollected experience.

There exists, however, another type of remembered information that has nothing to do with facts or events but relates to acquired skills like riding a bicycle or playing tennis. This remains intact in H.M. More intriguing, H.M. does not remember that he possesses this knowledge and, in fact, may deny that he can do what he is asked. For instance, H.M. learned how to mirror draw—that is, draw recognizable pictures while guided only by the reflection of his hand seen in a mirror. And although with each attempt his performance improves, proving that his brain remembers, he has no conscious memory of his earlier efforts.

Another test of acquired skill involves the famous Tower of Hanoi puzzle. This consists of three pegs and five wooden blocks which must be moved from the leftmost "Start" peg to the rightmost "Finish" peg according to set rules (only one block can be moved at a time, and a larger block can never be placed on top of a smaller block). In order to solve the puzzle, the subject must shuttle the blocks back and forth using all three pegs. There is an optimal sequence of moves (31), which H.M. learned at the normal rate. On retesting, he readily solved the puzzle despite no recollection of having encountered it before. In short, H.M. has learned without remembering having learned the skills necessary to solve the puzzle in the optimal way. Even when tested a year later, H.M. gave a very impressive performance.

Such findings fly in the face of our usual ideas about knowledge. Indeed we speak here of a dissociation of knowledge within the brain. The memory system impaired in H.M. is responsible for explicit memory (event and fact learning), that is, learning with awareness. The memory system preserved in H.M. mediates implicit memory (skill learning), that is, learning without awareness. Another way of describing the distinction comes from the philosopher Gilbert Ryle who distinguished between conscious directed attention to the act of remembering ("knowing that") and the knowledge required to perform motor skills ("knowing how"). Different parts of the brain are involved in these two types of memory.

In H.M.'s case, he suffered loss of the anterior medial part of both temporal lobes, including the temporal pole, the amygdaloid complex, and the anterior two thirds of the hippocampus. In terms of H.M.'s everyday living, this translates into a profound loss of continuity from day to day or even within a single day. He doesn't know how old he is, what he had for lunch, the identity of the President of the United States, or even that his parents are now dead. If anyone mentions that H.M.'s mother is no longer alive (she died just a few years ago), he becomes upset and grieves as if hearing the news for the first time. Despite this pure amnesia, H.M. is not intellectually deteriorated. He can conduct a perfectly normal conversation as long as he isn't asked to recall anything that has happened since his ill-fated operation.

In April 1993 Suzanne Corkin, the psychologist who has tested H.M. over the past thirty years, played a tape recording for me of a recent discussion with H.M. On the tape they exchange pleasantries and banter without anything odd or unusual occurring, until Corkin begins asking questions. H.M. does not recall what he did earlier in the day, any of the people who spoke with him, not even Corkin's identity. When she asks if they have met before, H.M. replies in a low sad voice and ventures that the two of them went to high school together. His voice, tentative and hesitant, suggests that he isn't very convinced about these assertions and, finally, in what sounds like a fatigued resignation, he admits that his memory is severely deficient.

H.M.'s severe impairment of explicit memory despite normal implicit memory skills suggests to memory researchers like neuropsychologist Mortimer Mishkin that the brain contains two interacting yet separate channels for memory storage. The first consists of the temporal lobe and its connections to the limbic system (responsible for specific recall: "knowing that"); the second is a more diffuse system located beneath the cerebral cortex in the area called the striatum and responsible for habit formation ("knowing how"). Other memory researchers believe Mishkin's two-part division an oversimplification and favor other divisions (an example in a moment). They point out that, in addition, the two pathways actually influence each other both directly and through intervening connections to other brain areas. But despite differences of opinion regarding the details all researchers agree that memory isn't at all the unity we have been led to believe on the basis of our subjective experience. Instead, it's a multicomponent process that can be broken down into distinct components and even subcomponents.

Nor is this fractionation of memory limited to people with brain damage. Normal volunteers who are asked to learn a series of inverted or otherwise transformed letters dramatically improve their performance with practice, and do so independently of their ability to remember the meaning of the sentences. First the brain analyzes information at the level of visual patterns. Analysis of content meanwhile proceeds on a different track. Pattern-analyzing experience is then stored separately from meaning. While the brain later recognizes the previous visual pattern and responds more quickly, the person exhibiting this enhanced performance responds like H.M. and has no recall of having encountered the patterns previously.

This division of memory within the brain into different tracks and subdivisions can be demonstrated experimentally. Since the experiment is a powerful example of applied neuroscience, let me say a few words about how it was carried out by a team from Yale University.

First, the researchers implanted microelectrodes into the prefrontal areas of a monkey's brain in order to record the activity of single neurons. In Chapter Eight we will have quite a bit more to say about the frontal and prefrontal cortex, but for now it's sufficient to know that the area is important in decision making and other executive functions. Specifically, the neurons in this area form a kind of time-binding function whereby representations—memories, if you prefer—are held "on line" for brief periods after the disappearance of the object in question. For example, if a peanut is shown to a monkey and then hidden under a cover, the monkey recalls the peanut's location thanks to the activity of the prefrontal cortex.

After placement of the microelectrodes, the monkeys learned two different tasks employing vision. In one, the monkeys learned to overcome their natural restlessness and sit and stare at a spot in the center of a video screen. Then, at a random moment, the experimenters briefly flashed an image to one of several locations on the screen. A few seconds later a cue on the screen prompted the monkeys to look back to where they had seen the image, a test of whether or not the monkey remembered its location. In a second test, the location of the image—a square—didn't change, but the pattern within the square did. The monkeys were trained to make no response until the image disappeared, and then look to the left if they had seen one pattern, to the right if they had seen another. This tested recall for features rather than location.

The Yale researchers discovered that neurons in one part of the prefrontal cortex respond to shape and color while another nearby but separate area responds to location. In other words, the division of labor for space and location established in the visual areas is continued forward into the most anterior and advanced portion of the brain. "Memory is modular; it's not all in one device," according to Patricia Goldman-Rakic, one of the Yale investigators. On the basis of this work, we have a "seeing-what" and a separate "seeing-where" modular organization, with the likelihood that each of the two systems can be enhanced or impaired.

These discoveries about the modular nature of memory stir up in me the following fantasy:

Imagine that as the result of some unspecified brain injury, you are amnesic for a two-week period in your life. During that period someone that you know well was murdered. You don't recall hearing anything about this, and when told, you respond with an appropriate grief reaction. But your claimed absence of memory raises the suspicions of the police who interrogate you about your whereabouts at the time of the killing. You of course can provide no alibi, since you don't remember anything from that period. You take a lie detector test and fail it. A witness appears who identifies you as a person seen with the victim several hours before the killing. The lawyer whom you hire in your defense knows nothing about amnesia and tells you he thinks your defense of amnesia is a weak one and that although the evidence is circumstantial you are likely to be indicted. At this point, how could *you* be sure that you hadn't committed the murder? I suspect you would fall back on self-assurances that murder is not something that you could ever imagine yourself doing. But how could you be really sure? In an attempt to demonstrate your innocence, you agree to taking another lie detector test which monitors your physiologic responses while you look at pictures, some of which contain items relevant to the murder scene. The results indicate an alerting response when you are presented with items related to the murder. Prior to your trial you read everything you can get your hands on concerning memory disorders, including the finding that recognition memory (your identifying the murder-related items) may, in the absence of conscious recall, be the only indication of prior experience. What would *now* form the basis for your conviction you had not killed your friend? Once again your conviction of your innocence could draw on nothing more substantial than your own sense of whether or not you considered yourself capable of murder. Memory would be no help here. And absent memory, the sense of identity crumbles (as with H.M.), since identity depends upon our ability to consciously bring to mind actions from our past. When we cannot do that, certainty about what we may have done in the past is impossible. In this hypothetical situation, you would be no more privy to the actual facts of the case than the jurors. If found guilty, you would no doubt protest your innocence. But a part of you would never be certain that you hadn't actually killed your friend during your amnesic period.

Loss of memory is so painful because it robs us of a part of ourselves. In an attempt to make up for that loss, some amnesiacs resort to confabulation: the production of "memories" which have no basis in actual events or occurrences. H.M.'s guess that he and his psychologist Suzanne Corkin had gone to high school together is an example. On occasion these pseudo-memories are so improbable as to earn a special designation: fantastic confabulation. For example, one seventy-two-year-old woman with a profound loss of memory told her doctor she had just come in from playing cricket, that she was a doctor in charge of an infirmary, and that her son was a field marshal. Usually amnesia sets off much less dramatic reactions. Patients of my own who have suffered concussions in auto accidents frequently voice distress at not being able to recall their accident. While such selective amnesia often seemed to me a merciful godsend for what surely must have been a frightening and unpleasant experience, the patients insisted on trying by various means (hypnosis, forced efforts at recall) to recover memories of their accidents. But their distress becomes understandable if we consider their memory loss as a gap in an otherwise intact sense of identity. We are ourselves, in fact can only be ourselves, by the exercise of our memory. But—as I hope I have convinced you by now—memory is not one thing but many, in fact is not a "thing" at all but an alliance of interacting systems with different functions. These systems operate throughout the brain in different ways with differing degrees of richness. The amnesiac or the Alzheimer patient is so disabled because he cannot, as the result of the memory loss, remember enough things from his past to sustain a sense of identity. If the illness is sufficiently severe, he may even lose the ability to recognize himself. Facial recognition is the correlate in the external world of the inner sense of identity. We recognize ourselves and those known to us because we remember having seen them before. But our memory for faces may undergo strange and puzzling transformations. Can you imagine what it might be like to look in your mirror and encounter a stranger looking back at you? Let us turn our attention now to some people who have undergone just such an alteration in their memory.

CHAPTER SEVEN

The Face That Launched
a Thousand Slips

A t a Christmas party I attended last year, my hostess said as she led me into her crowded living room, "Try to spend some time talking with the judge." Since I didn't know to whom she was referring, and didn't spy anyone in the room who looked even remotely like any judge I had ever encountered, I circulated and struck up several conversations with two or three of the guests. About an hour later, as I spoke to a pleasant-looking woman who introduced herself simply as Sandra, I realized I had seen this woman on television and pictured in newspapers and magazines. As our conversation continued I remembered my hostess's remark and considered asking this woman if she knew the identity of the judge. But, for reasons I couldn't articulate, this conversational gambit seemed ill-advised. Moments later I recognized my conversant: Supreme Court Justice Sandra Day O'Connor. But at what point did I become *aware* of who she was? The answer to that depends on our ideas of what it means to be aware.

As I entered the room, my past experiences of dealing only with male judges precluded me from even considering that the judge mentioned by my hostess might be a woman. This unconscious sexism resulted in the creation of a semantic category which I used when scanning the room: judges are male, and no man here looks very likely to be a judge. At this point it seems fair to say I had no awareness at all of the presence of the "Judge" Sandra Day O'Connor at the Christmas party.

Did my awareness of the true situation begin with recognition that my conversational companion was a celebrity of some sort? Or when I remembered for the first time in an hour my hostess's remark about the judge? Or did it start with the ill-defined but definite feeling of embarrassment about asking, "By the way, I understand there is a judge here. Do you know who he is?" Or did awareness occur only when face and name recognition coincided?

Disturbances in facial recognition can occur at different levels. We may completely fail to recognize a familiar face and, if asked about it, mistakenly respond that the person is a stranger. Or, as I did at the party, we may recognize the face as familiar but fail to recall any other anchoring details about the person, such as name or occupation. With the third and most common breakdown in the process of facial recognition, we may recognize and recall many things about the person, but not their name. As with my Christmas party experience, facial recognition starts off with a sense of familiarity, followed by recognition of the face as belonging in a specific category, and, finally, recall of the person's name.

In most instances, recognition of a familiar face occurs automatically. In fact, we cannot decide not to recognize a familiar face, although we are of course free to *pretend* otherwise. But I would suggest that, as with my Christmas party experience, recognition can occur in the absence of awareness, which may come much later or sometimes not at all.

With brain damage, the loss of facial recognition is exaggerated and takes a specific form called *prosopagnosia*: loss of the ability to recognize the faces of familiar people, even though recognition occurs immediately if the person speaks or moves or provides some other nonfacial clue. Vision, along with general intellectual ability, is perfectly normal, sometimes even superior, in persons with this disorder. Yet even the faces of family members, friends, famous people, and on occasion the person's own face in a mirror, go unrecognized. One patient with prosopagnosia when shown his wedding picture said: "Two people . . . one of them could be my wife because of the silhouette. . . . If it is my wife, the other person could be me." Another patient who could not recognize his wife or daughters if they approached him without speaking, once asked his wife. "Are you . . . ? I guess you are my wife because there are no other women at home, but I want to be reassured." It's important to note that in most instances of prosopagnosia, the afflicted person can still correctly interpret emotional facial expressions like happiness or anger, either genuine or mimed, and identify visual objects other than faces. The disorder is not a generalized difficulty with faces, rather the problem involves facial *recognition*.

Since the ability to recognize people from their faces plays such an important part in our lives, it should come as no surprise to learn that a specific part of the brain, the posterior right hemisphere, is involved in facial recognition. Prosopagnosia (literally "loss of knowledge of faces") usually results from a brain lesion in this area. But our interest in this strange disorder has less to do with where the causative injury is located than with what it tells us about human consciousness and awareness.

In 1983 a patient afflicted with prosopagnosia was given a variant of the Guilty Knowledge Test, a measure often used in criminal investigations. It is based on the frequent finding that a guilty person who denies committing a crime may often show some involuntary physiological response to questions or other references to the crime. For instance, if shown a picture of a murder victim, the killer may deny ever seeing the person before even though changes in his heart rate, breathing, or sweating responses indicate otherwise (I mentioned such tests in the previous chapter in the fantasy situation).

When the patient with prosopagnosia looked at photographs of familiar faces and tried to match names to the faces, he scored no better than he would have by blind guessing. Yet skin conductance responses, a physiological measurement, occurred maximally to the correct name 61 percent of the time. In other words, *overt* recognition of the faces, as measured by the patient's answers, lagged far behind *covert* recognition as measured by body response. Loss of *awareness* of recognition, rather than a simple failure of recognition, was the basis of his disorder.

Covert facial recognition in the absence of awareness also shows up in indirect measurements. The prosopagnosic patient often can match a face with its correct name faster than matching it to an incorrect one, for instance, indicating awareness in the absence of recognition (the patient remains unable to identify any of the faces when asked about them directly).

In most instances, patients afflicted with failures of facial recognition are concerned and frustrated by their failures. Their emotional response reflects their recognition difficulties even though covert recognition may be experimentally demonstrable. But occasionally a patient shows a double unawareness: He or she not only fails to recognize faces but is unaware of the facial recognition problem.

S.P. is an amateur artist who suffered brain injury from a burst aneurysm. She fails to recognize familiar faces despite retained ability to identify people by name and voice. Rather than experiencing distress or anxiety over her incapacities, S.P. maintains that she recognizes faces "as well as before." When her errors in facial recognition are tactfully but firmly pointed out by her neuropsychologist, Andrew W. Young, or a member of his team at the department of psychology, Durham University, Durham, England, she answers, according to Young, " . . . in a polite, somewhat detached, matter-of-fact way, with occasional expressions of mild surprise or disbelief." If the examiner persists, she offers the suggestion that the photograph is "a poor likeness" or that she had "no recollection of having seen that person before." When asked to identify portraits she painted previous to her brain injury, she does well, but only by means of an elaborate scheme involving skillful deduction from the sitter's apparent age, sex, dress, and other details. When asked about her failure to spontaneously recognize her own paintings, she insists that she had indeed "recognized them" and sees nothing unusual about the convoluted scheme she uses for identification.

S.P. experiences these difficulties because her perceptions of the faces she encounters lack any sense of familiarity. Absent insight into her impairment, she cannot monitor and correct her mistakes. She denies that anything is wrong, makes up excuses to cover herself, and concludes that what should be a familiar face is that of a stranger.

As with other neuropsychiatric impairments we are encountering in our study of the modular brain, S.P. suffers from an illness that attacks the very core of her being. Her impairment is similar to something like the loss of the ability to appreciate color, thereby restricting every perception to shades of black and white. Indeed her impairment is even more basic, somewhat akin to losing a primary component of psychological life. To her, facial recognition is not just "missing" but outside the boundaries of imagined experience. Her loss is like what would happen if a person were to lose all sense of the vertical or any concept of the past. And her lack of awareness of her impediment does not involve any change in her level of consciousness, intelligence, or other mental or emotional category of experience. Rather, something we all take for granted as part of our mental lives has simply ceased to exist for her. We speak here again of the existential dimension of human reality.

In normal people the feeling of familiarity usually leads to detection and correction of facial recognition errors. We incorrectly identify someone or fail to recognize them, but quickly correct ourselves. But this process can break down and any one of us, under circumstances of momentary uncertainty, can lose our sense of familiarity and, as a result, misidentify a familiar face.

In a now famous experiment, the psychologist D. M. Thompson convinced an Australian student to stand outside the London hotel where her parents were staying. Since the parents believed their daughter was home in Australia, they responded with pleasure and surprise when they saw her. But the daughter had been instructed by Thompson not to respond to her parents' greetings. Flustered, her father doubted his own powers of facial recognition and sense of familiarity. "I am terribly sorry, I thought you were someone else," he blurted out.

Facial recognition starts with a decision regarding familiarity (Do I know this person?). As with my cocktail party experience, we may initially fail to recognize someone due to preconceptions, inattention, or distraction. Facial recognition plays no part here. But when we come "face to face" with someone, our facial recognition system comes to the test. We can both recognize someone as familiar and at the same time *know* that we recognize them; or our consciousness may split: We have the "feeling" the person is familiar, the first stage of recognition, but remain unable to identify her. Yet measures of our bodily responses may indicate covert recognition: We know but don't *know that we know*. What happens next after our subtle, vague "feeling" of familiarity is often affected by the response that we get. As in Professor Thompson's experiment, even those with normally functioning facial recognition systems can be made to doubt the correctness of their recognition even of close relatives or intimate friends. It is as if the brain makes a best guess estimate of the other person's identity and then scans for indications of confirmation from the other person. If that confirmation is not forthcoming, doubts, confusion, and a sense of unreality may arise.

For example, one evening while exiting a cab in Cambridge, Massachusetts, I recognized a man I had known briefly ten years earlier in California. I hurried up to him with my wife trailing after me, put out my hand in greeting, and spoke my name in anticipation that he had forgotten it since last we met. Instead of taking my hand he smiled in a friendly but detached way, responded, "I think you have me confused with somebody else," and walked on. My embarrassment quickly gave way to a kind of panic, followed by a sense of unreality since I felt *certain* of his identity. But if the man really was a stranger, how could I feel such certainty? Vague notions concerning incipient insanity, instability, perhaps some form of brain disease went through my mind. At this point I felt I had no choice but to try again. Against my wife's advice, I ran up to him once more and said, "Surely you remember me . . . we were National Endowment of the Humanities Fellows at Stanford ten years ago." At this point his facial expression and manner changed dramatically at the recognition of who I was. Now that I had supplied him with the context, his facial recognition system came into play. Minutes later I felt comfortably in synch with myself once again, my identity restored, as the three of us sat in a bar and reminisced about our time at Stanford.

A similar momentary loss of one's sense of self is responsible, I believe, for the distress most people experience in response to a deliberate snub. More than just pride is involved when we are deliberately ignored. Rather, our sense of *identity* is attacked. Our recognition of the other person, gleaned principally from scanning their face, is not confirmed and we are treated as if we didn't exist. Indeed, in a way we don't exist. In the space of an instant our reality, our everyday sense of who we are and our relationship to the world, seems momentarily to disappear.

Thus facial recognition involves not only consciousness and awareness but individual identity as well. And although distortions and failures can occur secondary to injuries in a specific brain area, it's not really correct to think of prosopagnosia simply as a failure in the recognition of faces. S.P.'s disorder in awareness is more than simply a failure to employ a normal talent, facial recognition, but involves, according to Andrew Young, the loss of " . . . all knowledge of what it was like to have the relevant ability." In response to the social isolation and embarrassment resulting from the loss of facial recognition, bizarre and quirky rituals are oftentimes resorted to.

A medical doctor, Dr. S., wrote of her lifelong failures of facial recognition:

> I warn people now. If we have a wonderful exchange and are going to be friends for life, I say, Look, if we meet again outside . . . and I wouldn't recognize you, just give me the code word. The code word is something we talked about and then I would recall what we talked of.
>
> People complain I don't greet them. So I warn them beforehand. It is very embarrassing. I live in an active community . . . and I ask now

after four years in the community, "Who is this?" Now people cannot understand it. I ask friends of mine who do not know I have this problem, "Who is this?" They are amazed. "Of course, she is so and so."

Facial recognition is only the most dramatic example of a failure involving subjective experience. Other failures of memory—and memory does seem to be playing a large part in prosopagnosia—involve just the opposite conviction: memory illusions involving events that never took place.

All of us experience at some time or another the *déjà vu* feeling of familiarity when encountering something or someone for the first time. From here it is only a step toward believing, against all arguments by others to the contrary, that what we recall so clearly never actually took place. I am not speaking here of delusions or some rare or subtle mental disturbance. Neuropsychologists elicit it all the time in "normals."

In one demonstration, a series of words are flashed on a screen and the subject is asked to identify any words encountered on previous testing. Actually none of the words occurred previously; still, the subject can be made to select any word of the scientists choosing. There is a trick involved, of course. The word to be remembered is flashed on the screen moments before the start of the test at a speed too fast for the subject to consciously perceive it. Nevertheless, the flash presentation creates a sense of recognition. The word isn't actually remembered, but it does seem "familiar." The same thing happens if one word in a list is printed in more legible type than the others. The illusion of previously having seen the word in this instance arises because, unknowingly and for reasons I cannot explain, we tend to judge words printed with greater clarity than others as having been encountered before. In this case our brain correctly perceives that the word is somehow different from the others, but visual clarity is incorrectly assumed to result not from *visual* differences but from previous experience with the word.

Whenever I think about experiments like these, I begin to doubt the dependability of my own memories. If the sense of familiarity can provide us with firm convictions that things occurred that never actually happened, what hope is there that we can ever fulfill that Socratic dictum "Know thyself"?

As an example of my point, consider this test carried out by Larry L. Jacoby of McMaster in Ontario, Canada. Normal subjects read over a list containing the names of famous and nonfamous people. Later these names were included in a new list, and the subjects were asked to make a "fame judgment." All of the old names, whether that of a famous person or not, were more likely than any of the new names to be judged famous. Once again, familiarity leads to a false, totally unrelated attribution. Nor does it seem to make any difference whether or not the subjects consciously recall previously seeing the names (they were never asked to say whether a name had been seen earlier). Patients with severe amnesia who cannot consciously remember previously encountered names also exhibit the "false fame effect." Jacoby's interpretation of his experiments is so concise that I can't hope to improve upon it and therefore offer it in his own words:

> Memory can automatically influence the interpretation of later events. Those automatic influences of memory can be misattributed to sources other than the past and thereby produce a misleading subjective experience.

(I would only add here the reminder that the "memory" need not be a conscious one, and in many instances, as with the amnesiacs, the prior experience is totally inaccessible.)

> Effects of memory on subjective experience are automatic in that they do not require intent to use memory.

Jacoby's findings contradict another of our firmly established beliefs about ourselves, namely that we either recall something from the past and can with varying degrees of effort dredge it up, or we have forgotten it altogether and no amount of mental spadework is going to retrieve it. Instead it seems that we not only remember more than we know (supporting Freud's famous point about unconscious memories), but what we can't consciously recall may trip us up. Further, and this is the sticking point, I believe, previous events that we cannot recall may continue to exert powerful and unsuspected effects on our mental lives. Furthermore, our access to these events may remain extremely limited. I am thinking now of the claims, increasingly heard these days, that a memory of an early sexual seduction remained inaccessible until recovered during psychotherapy sessions aimed at "uncovering" childhood sexual abuse. Although many of these claims are based on fact, many others have proven to have resulted from suggestion on the part of the therapist. Memory expert Elizabeth F. Loftus, professor of psychology and adjunct professor of law at the University of Washington, has concluded after a lifetime of research and experimentation on human memory that " . . . not a single piece of empirical work in human memory supports the authenticity of claims of memory dating back to six months of age and research on later childhood amnesia raises many questions about repressed memory too." She points to over sixty years of research which has "found no controlled studies showing an event can be accurately reproduced in memory after a long period of repression." My point here is not to immerse ourselves in the increasingly rancorous debate about the truth or falsity of childhood sexual seduction. (Some cases undoubtedly occurred, while others are at least questionable; the difficulty is of course distinguishing the real from false memories of seductions that never took place.) Rather, I want to impress upon you the intimate connection that exists between memory and our sense of personal identity. We are what we remember, and in some cases we create a new identity for ourselves by remembering things from our past, in some cases things that never occurred.

If I could risk a one-sentence summary of what we have covered so far it would be this: Our subjective experience of remembering something does not imply that any such event ever took place, nor even that a specific "memory" for this supposed event exists in our minds. We are highly suggestible creatures, born storytellers, who possess a brain that is pre-wired to come up with explanations that provide us with a sense of meaningfulness. Gazzaniga speaks of the "interpreter," while Ken Heilman postulates within our brain the existence of a "comparator," which links past, present, and future events. Such a link is made possible by one especially important part of the brain: the frontal lobes.

CHAPTER EIGHT

Becky and Lucy

Martha Stoner, a successful businesswoman, at age forty-five experienced her first epileptic seizure. A CAT scan revealed a brain tumor in the right frontal lobe. After total surgical removal of the tumor, the seizures stopped and she recuperated at home for a year before returning to work. At a checkup several months later she reported no problems. Her husband, however, told a different story. Formerly an early riser, she now had difficulty getting up on time, showering, dressing, and getting to the office before 11:00 A.M. While at work she was easily distracted and shifted from one project to another without completing any of them. Her home and work behavior had deteriorated to the extent that her husband as well as her employer described her as a different individual.

Detailed psychological tests turned up no abnormalities. Her IQ remained above average and she showed no confusion, depression, or excitement. Most intriguing was the pattern of her difficulties. Her greatest problems involved organizing, scheduling, and sticking with tasks. She knew what had to be done, voiced a willingness to work, and yet could not coordinate the necessary steps to achieve what she wanted or what others asked of her. Rehabilitation exercises were started aimed at helping her focus attention, monitor her performance, and generally increase her awareness of how she was doing. Over eighteen months she failed to improve.

At this point she was asked to engage in a role-playing exercise in which she would evaluate her own work performance and make recommendations about what should be done. Her conclusions coincided with that of her supervisors: retirement. She accepted the decision. Outside of the role-playing situation, however, she was, in her doctor's words,

". . . no longer able to make the same judgments in relation to her real-life personal situation and was determined to return to work at her presurgery level of functioning."

At various places in this book we have encountered forms of denial which primarily involved the absence of *knowledge*. But Martha Stoner did not display a knowledge impairment. She knew full well what her problems were, but—and this is my main point—that knowledge wasn't any help to her in dealing with them. It was as if at one level she recognized her difficulties, while at another level she denied them—an especially curious combination in light of her normal to superior performance on neuropsychological tests, including memory.

Another peculiar aspect of her behavior: She had no problem with routine tasks, but failed miserably when forced to come up with a novel response. She was at her worst when called upon to focus her attention, abstract necessary information from background material, produce plans of action, respond with flexibility to changing circumstances, or evaluate the outcomes of her actions. These operations are carried out by the frontal lobes, the brain area destroyed by her tumor.

In comparison to other animals, the human brain has strikingly enlarged frontal lobes. Located at the front of the two cerebral hemispheres and encompassing almost 40 percent of the total cortical area of the brain, the frontal lobes have undergone greater expansion during evolution than any other brain area (they are 200 percent larger in humans than in other primates). Moreover, the most anterior part of the frontal lobes, the prefrontal areas located farthest forward, are so uniquely human that animal experiments offer only limited insight into human frontal lobe activities. For one thing, the frontal lobes are the last cortical areas to mature, with continued development through adolescence and into adulthood. In addition, the frontal lobes are connected to almost every other part of the brain, including the limbic system which is responsible for the experience and expression of emotion. These widespread connections form the basis for the frontal lobe's importance as an integrator and regulator of brain function. Thus a wide variety of mental processes may be interfered with by frontal disturbances, depending on the location of the damage.

At one time the frontal lobes were known as the "silent areas" of the brain. That was based on their lack of response to electrical stimulation. In other parts of the brain an applied electrical current usually produced some measurable or observable response. Depending on the site chosen for stimulation, a limb would jerk outward, or a temporary paralysis would ensue, or, if the "test animal" was a human, he or she might describe flashing lights. But nothing like this accompanied frontal lobe stimulation. If the frontal lobes carried a special function, it was certainly not an obvious one.

In the 1870s a pair of monkeys surgically deprived of their frontal lobes were described by the experimenter as exhibiting "a decided alteration in the animal's character and behavior," but one which was "difficult to describe precisely." To the casual observer the monkeys appeared no different than before the operation, since they exhibited no defects in sensation, perception, or motor power. But closer observation revealed the monkeys remained largely apathetic and uninterested in their surroundings. ". . . while not absolutely demented, they had lost, to all appearances, the faculty of attention and intelligent observation."

Other monkeys and several dogs subjected to the same operation also displayed a profound change in personality. Fear replaced their usual affection for their owners and caretakers. They stopped playing, socializing, and grooming. If frustrated, they reacted with violence. The researcher, Leonard Bianchi, an Italian neuropsychiatrist, explained the animals' behavior as a "disaggregation of the personality" and blamed the animals' change in behavior on their failure to integrate experiences.

Alerted to the possible importance of the frontal lobes, neurologists soon found human correlates of Bianchi's personality "disaggregation." One of the most carefully investigated was the patient of neurologist Leonore Welt. This thirty-seven-year-old man fell from a fourth-floor window and sustained a penetrating fracture of the frontal bone. His *physical* recovery followed the removal of the bone from the frontal lobe. But *mentally* and *emotionally* he was a changed man. Formerly a hard-working, successful furrier, he lost his ability to contain his temper and flew into rages at the slightest provocation. His manner was teasing and insolent to other patients, his doctor, and the nurses. But within a month his behavior gradually improved. He stopped quarreling and expressed remorse and embarrassment over his prior behavior. Several months after discharge from the hospital and his return to business

(Welt made no comments about how he did at work), the man died of pneumonia. Autopsy revealed destruction of basal portions of both frontal lobes.

Welt's patient, along with others with similar injuries, provided proof that the frontal areas are not "silent." A patient (identified simply as A.) had both frontal lobes removed by the famed neurosurgeon Walter Dandy in order to prevent the spread of a brain tumor. His postoperative behavior was so dramatically altered that he became the subject of an entire book written by his neurologist (*The intellectual functions of the frontal lobes: Study based upon observations of a man after partial bilateral lobectomy*). Formerly hard-working and reserved in disposition, he turned boastful, uninhibited in speech, and unconcerned about his work, family, or future prospects. Yet his intellectual powers remained largely intact, as did his language, perception, and other cognitive processes. He was in essence a puzzlement to his doctors and a burden to his family and friends.

Additional light was shed upon the functions of the frontal lobes as a result of additional experiments on monkeys. Carlyle Jacobson of Yale University induced what he called an "experimental neurosis" in one animal by forcing her to perform experimental tasks too difficult for her to master. Frustrated, she flew into a rage and refused to make any further efforts. After the excision of both frontal areas, the animal was willing to undertake the tasks again but with an important change in behavior. She was no longer upset by failure and displayed no "temper tantrums."

Impressed with similar results in another female monkey (Jacobson affectionately referred to the pair as Becky and Lucy), the Yale researcher put on a sophisticated "show and tell" demonstration at the 1935 International Congress of Neurology in London. The demonstration that formerly fierce and aggressive animals could be turned docile by means of a fairly straightforward brain operation suggested to one of the attendees, the Portuguese neurologist and professional diplomat Egas Moniz, that the operation might prove helpful in controlling violent and assaultive mental patients. (Moniz would die years later at the hands of just such a patient.) At the conclusion of the congress, Moniz returned to Lisbon and initiated the most extensive and radical psychiatric "treatment" in medical history.

Psychosurgery would not have been possible except in an age notable for doctors daring and foolhardy enough to carry out on humans what was an essentially an animal experiment; and except for patients sufficiently devastated by their illness to silence any objections to the procedure, and family members desperate enough and gullible enough to try anything that might quiet disturbed and disturbing relatives. And quiet them the operation certainly did. But my purpose here is not to develop the story of psychosurgery in any great detail (I have done so in one of my previous books, *The Brain: The Last Frontier*), but to present some observations made by several of the key players active in psychosurgery.

"They [postlobotomy patients] see their faults and weaknesses and vices clearly enough and recognize them for what they are, but far from privately brooding over them, they seem to regard them as interesting topics for comment," wrote Walter Freeman, a neurologist and psychosurgery zealot who developed his own macabre office-based procedure utilizing an ice pick shoved up into the frontal lobes through the thin plate of bone lining the inner canthus of the eye.

Others spoke of frontal lobe destruction as producing "lack of concern," loss of the "abstract attitude," and a mental state marked by mood disorders, disturbances of attention, and a lack of drive or initiative. But however it is described, frontal lobe disease is a very debilitating illness that destroys or impairs those aspects of our thought and behavior that are most characteristic of our species. We are the only creatures capable of anticipating the future consequences of present actions; setting up plans and goals and working towards their achievement; balancing and controlling our emotions; and maintaining a sense of ourselves as active contributors toward our future well-being. These powers are diminished or lost with frontal lobe disease.

Two frontal lobe syndromes have been described, depending on the location of the damage. Injury to the orbital frontal cortex along the underside produces impulsivity, rapid mood changes, irritability, and sometimes physical aggression. If the injury is to the lateral parts of the frontal lobes, the person loses all initiative and spontaneity; the mood is flat and disengaged; initiative is replaced by inertia and a "couldn't care less" attitude. Although these two forms of frontal lobe disturbance seem quite different, they share a basic failure in regulating conduct and behavior in accordance with either social convention or internal goals.

Most neurologists who have studied the frontal lobe and its disturbances emphasize three critical functions: self-activation, resulting in a person changing his outer or inner world in some way; self-monitoring of ongoing behavior coupled with the capacity to alter it in response to positive or negative feedback; finally, planning and organizing behavior over time in the absence of cues either from other people or ongoing events. When functioning normally the frontal lobe makes possible the necessary flexibility for adapting to new challenges. The person afflicted with a frontal lobe disorder cannot do that, hence the two behavioral disorders which are two sides of the same coin: lack of initiative and apathy versus erratic impulsive actions.

But the one frontal lobe function I want to emphasize is what for a better term I call the "sense of self." I'm not referring here to self-consciousness or self-esteem, although both of these play a role. Rather, the "sense of self" places a major emphasis on our relationships with other people. A healthy "sense of self" involves a balance between seeing things from one's own personal and subjectively intimate point of view, while at the same time retaining the capacity to step back and imagine how others may view the same situation. We speak here, as we have at other points in our exploration, of *awareness*. But the awareness contributed by the frontal lobes involves more than simple denial. As with Martha Stoner, a person may speak openly and seemingly candidly about a deficiency and yet this "insight" exerts no effect on their present behavior.

A patient of mine suffered frontal lobe damage as the result of the rupture of a blood vessel. After surgery and rehabilitation, he returned to his job in personnel management. He had suffered no loss of mental abilities as a result of his illness and appeared motivated, even anxious to get back to work. After several weeks on the job he returned to me with a long letter from his supervisors detailing changes in his personality and behavior. Formerly a "self-starter," he now required almost constant supervision and had to be prodded into performing even routine office tasks. He showed no signs of resentment when confronted about his performance and always offered assurances that he would improve.

In fact, he saw nothing wrong with his performance, not even on the occasion during a staff meeting when he fell asleep. "I was only resting my eyes," he said. When I asked regarding his difficulties, he only admitted to being "a little slow" and expressed confidence he would soon "get up to speed."

When at home he spent almost all of his time watching television, rarely read anything, spoke little, and preferred staying inside rather than, as was his former wont, visiting with friends. Again, he claimed that although this represented a deviation from his former behavior, he felt much the same about his work, wife, and friends.

Psychological tests were not helpful in revealing Robert's problem. "His performance on the neuropsychological screening, which included many of the tests considered most sensitive to the presence of cortical dysfunction, was entirely within normal limits," wrote the neuropsychologist in his report. Despite this conclusion, my review of the test materials turned up a pattern of forgetfulness following delay or distraction. He could quickly learn and recite a sixteen-item shopping list, for instance, but forgot most of the items after a short delay or if distracted and asked to turn his attention to something else. "He is highly dependent upon the presence of context in his ability to encode and store verbal information," according to the report.

Even more revealing of Robert's frontal lobe disturbance was this observation by his supervisor concerning his performance on a computer. "Unless he is told specifically what to enter into the systems, he seems to have a great deal of trouble focusing his attention on the job at hand. He is continually printing unnecessary copies of every screen he has on his terminal. He buries himself in paper and has trouble weeding out unnecessary data. He appears to have difficulty keeping sequential steps in order. Without external intervention, he leaps from task to task lacking a sense or understanding of purpose. He gets sidetracked by all the paper he is looking through."

Throughout all this Robert remained blissfully unaware of his failures. In those instances when denial was impossible, he said confidently that he would rapidly improve. But little or no improvement took place. In explanation, he said all of the components for normal performance were present but he had "difficulty getting everything together."

Robert illustrates the five major prefrontal functions suggested by frontal lobe specialists neurologist Frank Benson and neuropsychologist Donald T. Stuss.

- *Drive/Motivation.* Frontal lobe damage results in the loss of ambition and self-motivated behavior. External sources of stimulation and inspiration become more important as motivators than internal self-direction. The afflicted person is apathetic and, in the absence of constant supervision, monitoring, and external stimulation, does nothing other than sit staring into space.

- *Sequencing.* As with Robert at the computer, the frontal-lobe–damaged person lacks the ability to keep bits of information in proper sequence. He also has difficulty in separating the most essential information from less important background material. "Buried in paper," the afflicted person wastes much time in going back over the same material in an unsuccessful attempt to form the necessary integrations to reach a conclusion.

- *Executive Control.* Planning and anticipating the consequences of behavior are disrupted with frontal lobe damage. Robert could not monitor his own performance and therefore rigidly adhered to unsuccessful approaches. In some patients this loss of self-monitoring extends to social behavior. They may make crude and insulting comments, tell obscene stories, or openly express frank sexual interest only moments after meeting someone new. They exercise poor judgment and generally lack the ability to see things from other people's point of view.

- *Future memory.* Almost all complex human activities require some advance planning. This involves imaginatively comparing how things are with how one wishes them to be. The resulting internal model of the future serves as a guide for altering and updating one's behavior in the direction of achieving the goal. In this way, the individual is acting on what has been called a "memory of the future." With frontal damage, stable internal models seem not to exist. Something as simple as phoning to make a train reservation is fractionated into separate components that occur out of sequence and even interfere with one another.

- *Self-Analysis.* Walter Freeman and James Watts, who between them carried out in the 1940s and 1950s more psychosurgical operations than anyone else in the United States, spoke of a disruption in the sense of self-continuity: recognition of a stable self that exists from the past through the present into the future. As the result of a lobotomy, the patient lost awareness of himself as a changing, evolving person with personal responsibility for the changes affecting him. He could no longer project himself into the future or integrate experiences from the past. Lost, according to Freeman and Watts, was ". . . an intimate awareness of the relatedness of what the self has done and experienced to what it will do and experience." And the importance of this loss stems from the fact that "Self-awareness, consciousness, or self-reflectiveness is the highest psychological attribute of the frontal lobes," according to psychologist Donald T. Stuss.

Specific personality responses ensue from these defects. The afflicted person is often tactless, querulous, argumentative, and given to angry outbursts when challenged. Impatient of restraint, and with a lowered tolerance for frustration, noise, or bustle, his impatience can quickly escalate from angry words to outright physical aggression. These responses can erupt from a personality marked either by depression and withdrawal, or hyperactivity and a surface extroversion and even euphoria.

In illustration of the role of the frontal lobe in normal mental processing, Jordan Grafman, chief of the Cognitive Neuroscience Section at the National Institutes of Health, would have us imagine a bank manager sitting at his desk. Before him he sees pens and pencils, a writing tablet containing notes from yesterday's staff meeting, a calendar with reminders of future engagements, and a computer which contains a software program explaining the new banking regulations. In the background soft mood music can be heard, along with intermittent street noises.

When he focuses his attention, the banker must be able to recognize the various objects on his desk and their functions, and be aware of the reason he has scheduled meetings in the future, and of the relevance of these future meetings to the general content of yesterday's meeting. Picking up his notebook or typing on his computer terminal, the manager composes a memorandum for himself, along with an interoffice memo. While writing, he finds it necessary to exert a firm mental effort in order to focus his attention away from the intruding but pleasurable fantasy of his upcoming trip to the Caribbean.

This everyday scenario illustrates some of the obvious demands made on the brain when it processes information. Objects must be recognized and distinguished from others; events in the past must be clearly distinguished from the present and related to the future; things must be put into context; attention must be focused against the intrusion of external events and the distraction of internal thoughts, impulses, and fantasies.

In addition, Grafman suggests there are other less obvious mental processes going on, processes the bank manager is unlikely to think consciously about. How does he know where he is, for instance? How does he know what he is supposed to be doing while at his desk? If someone comes into his office, how does he decide what is appropriate behavior under the circumstances? When he answers the telephone, how does he decide what to say and in what tone of voice and with what manner?

Our hypothetical bank manager knows what to do because over his lifetime he has stored in his brain many behavioral sequences Grafman refers to as Managerial Knowledge Units (MKUs). These sequences involve events, real or imaginary (a daydream), that have a beginning and an end and are retrieved automatically. Engaging in a conversation or eating a meal at a restaurant, or, as with the bank manager, working through a typical "day," are examples of MKUs. Some of these activities take only a few moments, others may stretch out over hours or even the whole day. This provides a challenge for the brain, since working memory allows for only a time frame of about ten seconds. As the duration of an activity exceeds the window of working memory (e.g., as with our bank manager, who may require several hours in order to complete his memorandum), the activity represents simultaneously the present (writing the memo), the past (thinking about what he wishes to say), and the future (anticipating the responses the memo is likely to evoke). The manager must, in essence, hold on-line, to borrow a computer term, events from past and present while linking them with the future. "The ability to project the results of current actions into the future and to reason about the time course and duration of the resulting events is not a minor development but a major evolutionary step," says Grafman. What would things be like if our species had not made this evolutionary step? Grafman suggests that just such an impairment exists following damage to the frontal lobes of the brain.

Time-locked in the present and lacking the ability to anticipate the future or link the present with the past, the frontally injured person is missing a critical human faculty: personal continuity resulting from sustained attention and awareness over time. Without this faculty, long-term planning becomes impossible.

For Grafman, our prior experiences form the foundation for future "willed" action. We can only do what we can imagine doing, and that depends upon our stored memories derived from our own experience or our knowledge of somebody else's experience. To this extent—and Grafman freely admits it—his emphasis is on the environment, specifically the events in our past, now stored in memory, that provide us with models of how things "are in the world." Frontal lobe disease interferes with this process, in many instances disrupts it altogether. The affected person loses the ability to initiate a plan of action, stick with it, and reach a goal. Moreover, no generally available methods so far have proven successful in assisting frontal lobe patients with their planning difficulties. As with my patient Robert, they keep saying they will do better but prove unable to sustain their early efforts.

Grafman's ideas were developed on the basis of an extensive study of frontal lobe injuries sustained by soldiers in the Vietnam war. They provide good reason to doubt that we humans ever act completely freely. Indeed, this notion that we are unconstrained in our capacities to act—what Jordan Grafman refers to as the "romantic notion" of free will—ignores the fact that except for some few extraordinarily creative individuals, most people's actions are based upon the activation of stored memories in the form of scripts, plans, schemas—call them what you will—that provide models for what *might* be done under the circumstances of any given moment. Indeed, human actions depend upon their activation, largely outside of our conscious awareness, over our lifetime.

"Our choices are finite and based upon our stored memories. These memories may even be fantasies or internal representations of the world that have never actually been acted out. But in any case our so-called free will comes with some constraints placed upon it," says Grafman.

* * *

Frontal lobe disease gets us to the heart of our search for the neurological underpinning of the personal self. Deprived of a sense of continuity, the affected person remains isolated and buffeted by the winds of chance and unpredictability. Integration and stability disappear. Existing wholly in the present, the future cannot be envisioned nor the past meaningfully recalled. And this loss differs from the memory disturbances discussed in the last chapter. Rather than memories of fact, we speak here of subjective, personal, *experienced* memories. The psychologist William James gives the example of two people waking up in the same bed. Even though each may have some intuition, based on the last things discussed the night before, about the other's thoughts upon awakening, direct access is only possible to one's own thoughts. These are marked by a sense of immediacy, warmth, and intimacy. "The universal conscious fact is not 'feelings and thought exist' but 'I think' and 'I feel,'" as James put it.

Deprived of the sense of personal integration and the "ownership" of his own mental activities, the person with frontal lobe damage inhabits a robotic world. It would not be an exaggeration to say he is deprived of his humanity, his identity as a member of the human community. Further, this deprivation progresses along a continuum. Although its beginnings are often subtle, progressive frontal lobe deterioration inevitably results in a sadly pathetic caricature of the human personality. Here is a description of advanced frontal lobe disease that captures the manifestations of the illness in its advanced stages:

"It is marked by personality change and breakdown in social behavior. Patients become unconcerned and lacking in initiative, and they neglect personal responsibilities, leading to mismanagement of domestic and financial affairs and impaired occupational performance. Medical referral may occur following demotion or dismissal from work. Affect is invariably shallow and emotional empathy with others is lost. Rigidity and inflexibility of thinking and impaired judgment are characteristic. Patients vary, however, with respect to certain behavioral features. Some patients present as overactive, restless, highly distractable, and overtly disinhibited. Their affect may appear fatuous and superficially jocular. . . . Other patients present with apathy, inertia, aspontaneity, and emotional blunting."

In earlier times such persons would have been described as having suffered a loss of the soul. And this seems not too much of an exaggeration, since the illness deprives its victims of those aspects of themselves that distinguishes them from brutes. Also lost are the individuating qualities that distinguish them from other people. In place of the novelty and variation of a individual personality, the frontal-lobe–injured person displays a frightful sameness and invariability.

How curious and sobering it is to realize that our most advanced and evolved mental activities depend on unimpaired functioning of a specific part of the brain. As another way of putting it, our most human traits exist for us as a function of the human brain. Further, damage to our frontal areas could reduce any of us to an almost subhuman level of functioning, a kind of psychic limbo where we dwell in an eternal present, devoid of what I consider our most evolved mental ability: our capacity to empathize with others. No other creature, including the higher primates, comforts the injured or the bereaved, because other creatures cannot imaginatively identify with another. Prehistoric man was also singularly lacking in this capacity to put himself in another's place. With the development of this capacity within our species, thanks to the growth and enhanced functioning of the frontal lobes, came the capacity for imaginative identification with others. Societies and organized communities followed. Not all of them evidenced frontal-lobe–mediated qualities, and in the absence of some of the more important ones, like anticipating the future consequences of present policies, some of them perished. Can contemporary societies, as with the individuals who comprise them, exhibit variations in the experience and expression of frontal lobe function?

A high school teacher in the South Bronx wrote me the following letter after attending my lecture on the frontal lobes:

". . . You mentioned during your lecture [that] 'The psychological functions of attention, motivation, and emotion, like autonomy, are affected by frontal lobe damage. The overall result is a lessening of the individual's free will.' In dealing with a high school population of profoundly underachieving adolescent/young adult males between the ages of 14 and 19, we regularly witness students whose attention span, motivation, autonomy, and emotion appear to be diminished, if not restricted. Many of these kids lack age appropriate cognitive development as demonstrated by their lack of insight, abstraction, concept formation, inferential attitude, organization and planning, and a sense of autonomy. . . . Now my question is: Could environmental influences, i.e., the domestic violence, economic and social deprivations that exist here in the South Bronx, result in these subtle frontal influences you spoke about?"

This question, intriguing but unanswerable, rests upon an intuitive grasp of the interlinking influences of the brain and the world in which it finds itself. The writer continues: ". . . Is it possible that many of these children are being deprived directly or indirectly of adequate frontal development because of environmental deficiencies? After all, trauma could be covert as well as overt; psychosocial as well as physical, could they not?" As this dedicated and concerned teacher has observed, frontal lobe characteristics, particularly empathy, vary widely within different segments of our culture. The brutal and treacherous conditions existing in our inner cities almost guarantees that its inhabitants will be preoccupied with simple survival. There is little time or place for the development of such frontal-mediated qualities as ethicality, compassion, and planning for the future.

Although he never spoke specifically of the frontal lobes, psychologist Abraham Maslow grasped this dilemma many years ago when he pointed out that human motivation consists of a hierarchy ascending from basic biological needs (hunger, thirst, safety) to more complex psychological motives: self-esteem, the cognitive needs to know, understand, and explore; the aesthetic need for order and beauty; and finally, at the pinnacle of the hierarchy of needs, the need for self-actualization and the realization of one's identity and potential. The higher needs mediated by the frontal lobes are fulfilled only after the satisfaction of more basic biological needs. And no reasonable doubt exists that large portions of our population know only the reality related to the most basic of needs. One has only to read through the metro sections of the newspapers in any of our major cities to encounter horrifying tales of brutality and viciousness. Empathy for others is not a universal trait despite the fact that all us have the potential to be empathic.

We have learned that our knowledge of the frontal lobes provides a new way of looking at important and puzzling social dysfunction. If the frontal lobes are not nurtured and developed in portions of our population, perhaps as a result of the factors mentioned in the teacher's letter, then we as a society can expect to continue to pay dearly in terms of more crime, broken homes, drug use, and violence. At the basis of each of these disorders lies a lack or distortion of our natural human desire for freedom. We achieve this freedom via the exercise of free will. But do we possess free will or are we determined by our environment? Although such a question seems to bear little relation to neuroscience, but seems of more concern to philosophy—indeed it has fueled all of Western philosophy since the Greeks—considerable light has been shed on the topic by recent discoveries about the human brain.

CHAPTER NINE

The Alien Hand

I n France after the Revolution a philosopher with the mellifluous name Pierre-François Gonthier de Biran set out to write a monumental work he entitled *The Science of Man*. He never completed it, because of a decision during his later years (he lived from 1776 to 1824) to devote himself to understanding the nature of willed action. Biran considered will as the "primitive fact of psychology which sheds light on every other aspect of the human mind."

Biran started out with the observation that the operation of the will could only be known through immediate subjective experience, a feeling of effort exerted against some external object or force. But he broke new ground with the claim that the operation of the will could not be broken down into two operations, as it had been traditionally. Thus, when a person raises an arm, the action is willed but does not, Biran insisted, consist of an inner event—the decision to move the arm—followed by the motion of the arm. Instead, will consists of one unified and indissoluble action. Biran declared that the mind deals with facts and relations (the deliberate raising of the arm) and not just objects (the arm) acted on by some independent sense of effort. In addition, willing is inseparable from knowledge, Biran claimed, since a person cannot act with deliberation unless she knows what she is doing. Thus knowing and acting are reciprocal: to act is to know and vice versa. With this as his base, Biran went on to consider will, thought, consciousness, and knowledge as *active* processes of the human mind.

Biran is remembered today for his insight that the mind is an active participant in the creation of our "reality," not a passive recorder and responder to the world in which we find ourselves. But recent discoveries about the brain's operations give us reason to doubt the correctness of his insistence that will cannot exist independently of a specific willed action.

But before getting to that research, consider the following thought experiment. At the moment I am sitting at my desk which looks out on Vineyard Sound on the island of Martha's Vineyard. As I look out and catch a glimpse of a boat owned by a friend, I'm caught up in the fleeting impulse to run down to the dock in order to wave my friend in to pick me up to go with him on a sail. But I decide against this in order to stick with my writing. Yet while vetoing the run to the dock, I certainly feel personally and intimately connected with my willed decision to stay where I am. I also experience the pull of forces directing me to go sailing. In both instances, it seems to me, my will attaches itself to an internal representation, a picture if you prefer, of what it would be like to while away the afternoon on a sailboat versus another representation of what it would be like to not finish my book on schedule. Neither of these two possibilities seems more vividly represented in my imagination than the other, yet I choose one. Biran would remind me that my choosing is an active process and that I know perfectly well what I am doing: to know is to act, to act is to know. But my not going down to the dock also involved knowledge (the scenario of not completing the book) and an act as well: the act of *not* getting up from my chair, not leaving my word processor.

By tradition we in the Western world tend to equate will with doing. But sometimes the most incisive instances of will are notable for the absence of anything happening in the external world. A famous Zen story captures my meaning:

Two famous and experienced Samurai were pitted against each other in a fight to the death. Over the space of an hour they moved closer and closer together until less than an arm's length separated them. At this short distance they could study each other's pupillary reactions, breathing patterns, the occurrence of any momentary movements denoting fear. Soon another hour had passed without either of them attacking. They merely watched each other with full and attentive awareness. As one of the fighters considered an offensive action, the other recognized what was coming and somehow communicated without moving a muscle his perception and readiness to respond. Over the next two hours each potential attack, repulse, and counterattack was thus worked out in a purely mental exercise which remained invisible to all but the combatants. After a total of three hours of this the two combatants, without saying a word or employing any discernible communication between them, recognized that they were perfectly matched. They bowed solemnly to each other and at the same instant each backed away.

Not moving the arm was the willed act of the Samurai warriors, and it rested on an internal representation of the negative personal consequences of an attack on the opponent. Although it is not the preferred use of the term, our culture recognizes inhibitory instances of willed action. We describe some of our friends and acquaintances as "strong-willed" because they can stick with rigorous diets while all around them are eating to excess. We also apply terms like "weak-willed" to those who cannot resist temptations of one kind or another. And although such designations often attempt to disguise our bias against people we dislike, the attribution of a "strong" will to another does seem to have something to do with that person's ability to retain and direct a focused attention towards a selected goal. William James and other nineteenth-century psychologists recognized the value of this and wrote books and delivered speeches aimed at teachers and parents instructing them how they might foster "will power" in young children. But whatever the goals and methods employed to develop the will, we now recognize the underlying process involves brain programs which, as with apraxia, can go dreadfully awry. Once again we run up against the burning issue of identity: Who or what is doing this willing? "It is *I* who decides and wills . . . ," we are quick to respond. But how do we do it, and what parts of the brain are involved in the exercise of will?

Several years ago I was consulted about a young man who had suffered a stroke involving the more anterior parts of the brain, including part of the frontal lobes, which had left him able to move about normally enough but seemingly without any desire to speak. He would sit all day in the same place, and if spoken to would rarely if ever respond. On those occasions when he did speak, usually uttering nothing more than a single phrase, his speech was understandable and he made perfect sense. From this I concluded that the relevant brain areas for speech remained intact but unactivated for some reason. It was as if he lacked the inner drive necessary to activate the speech process. I decided to treat him with a high dose of the psychostimulant amphetamine. After the first dose, he began moving about and speaking spontaneously and in response to questions. He told me that after his stroke he had retained his desire to speak to others, do things, and move about generally, but couldn't "force" himself to get started. It was as if he had lost his will.

Another young man (not a patient of my own but one reported in the literature), tried killing himself by hooking up a hose from his car exhaust and breathing the carbon monoxide fumes. At the last minute a friend discovered him in the garage and summoned a rescue squad, which after resuscitation transferred him to a nearby hospital. A CAT scan showed damage to several nuclei beneath the cortex which ordinarily are involved in movement. After regaining consciousness two days later, the man remained completely immobile, as if frozen. In his case no drugs were required in order to arouse him from his mannikin-like state. The tactile element was crucial here. He could be made to move if gently prodded or poked; the lightest of physical contacts, and he would spring to life. At such times he reported that during his unresponsive periods he had been fully aware of what was going on around him and wanted desperately to respond when spoken to, but couldn't force himself to speak or act.

Less severe and less extensive damage to the will may take on more bizarre features. A patient afflicted with what neurologists call the *alien hand sign* may have no difficulty with initiating movement or speech but may declare instead that he has lost control over the hand on the opposite side from the brain damage. For example, the hand of one patient I encountered would on occasion reach across at the dinner table and take food from the other hand which pulled away in protest. When questioned about this strange behavior, the patient typically responded that the offending hand was outside of his control, that some other person or thing was moving it. He didn't deny that the hand belonged to him, as did Helen French whom we met in Chapter Four. Rather, he acknowledged the hand as his own but failed to experience himself as in control of it.

All three of these individuals—the two mute patients and the patient exhibiting the alien hand sign—suffered from disorders of the will. But despite the compromise of willed action in each instance, the areas of brain damage were not the same. This makes it unlikely that any specific area of the brain is responsible for will. But not all areas are of equal importance and some localization is possible. Neurologists have even come up with a rule that helps in understanding these strange matters: The more anterior the brain damage, the more likely the injured person will incur some loss of the feeling of volition and control. Recall Robert, the patient with frontal lobe lesions we discussed in Chapter Eight. His greatest difficulty involved self-motivation. But when we take Robert and other brain-damaged individuals as examples, we run into a conundrum: Can we extend to normals these findings about will garnered from individuals with will disorders secondary to brain damage?

In order to solve this, neuroscientists employed electrical recording techniques to measure what goes on in the brain during willed action. After the establishment of a conditioned response—like Pavlov's conditioning of his dogs to salivate at the sound of a bell—a special electrical potential can be recorded from the brain. This is how it works.

After conditioned learning to associate one stimulus with another, the animal (or human, since conditioning procedures are perfectly natural, such as our conditioning to respond to the sound of the alarm clock in the morning by heading to the kitchen for our coffee) is hooked up with recording electrodes. In a typical test, there is a period of expectation between the first stimulus and the arrival of the second stimulus a few seconds later. Just prior to this second stimulus comes an electrical discharge, the CNV (contingent negative variation), recorded maximally over the frontal areas. This response has nothing to do with whether the subject actually does what he or she was conditioned to do. The response is a measure of *expectancy*. This was the first measurement of something happening in the brain of a person or animal not engaged in some form of outward activity. Everything here is strictly "in the mind."

The second revolutionary discovery about the brain, the readiness potential or RP, was the work of a German neurologist, H. H. Kornhuber. This is a negative wave generated during the mere *intention* to bring about a brief voluntary action like raising a finger. The RP is maximal in the vicinity of the supplementary motor cortex and differs from the CNV mentioned a moment ago by virtue of the fact that no conditioning is involved and the subject is free to raise the finger on a whim. First occurs the RP, followed by the occurrence of the voluntary urge or intention.

Kornhuber's experiment would have intrigued Descartes, since it restores to its once eminent position the primacy of internal processes as the means of understanding the mind. Put somewhat differently, if Descartes had been hooked up to one of Kornhuber's machines, he would have been able to say with some justification, "See, I told you so."

More recently a neurophysiologist at the University of California in San Francisco has further elaborated on the discoveries of Kornhuber. In Benjamin Libet's experiments, his subjects are requested to make a simple flexion of their forefinger at a moment of their own choosing. He finds that at about 200 milliseconds (a fifth of a second) before the movement, the subjects decide when to make the movement. Libet knows this because his subjects relate the moment of their decision to the position of a dot of light revolving around the clockface. ("When I first made up my mind, the light was at such and such a position on the clock.") But 350 milliseconds (about a third of a second) before his subject's conscious decision, Libet records a sudden flurry of brain activity. Now that initial 200-millisecond delay is explained easily enough as the time it takes for the finger muscles to actually respond. But that 350-millisecond delay eludes explanation. As Libet explains it: "The brain 'decides' to initiate or at least to prepare to initiate the act before there is any reportable subjective awareness that such a decision has taken place."

Libet's findings provide a corrective for any hubris we might feel about our power of will. "This evidence certainly constrains the way we may view free will—as a mechanism of self-control but not as an initiator of the intention to act."

Kornhuber's and Libet's experiments, along with apraxic disturbances, run counter to our intuitions about how we go about performing voluntary actions. It is not a matter of simply deciding something and then *just doing* it. Breakdowns can occur at different points and as the result of damage in different brain areas. Further, our actions share many of the same processes as our speech (a portion of the supplementary area activates just prior to speech). Both occur in a temporal series. Just as we go about pouring ourselves a cocktail in an invariant sequence (taking the bottle from the shelf, then opening it, then pouring out the liquor, and so on), we must, in order to be understood, utter our words and the sounds comprising our words in a particular order and time sequence. Both speech and action require exquisite coordination of large numbers of muscles. Any alteration in sequence and we "stumble" over what we are saying or doing.

Whenever things go wrong in a sequence, the explanation usually invokes some type of disruption in a "motor program." Thus the apraxic failure mentioned several chapters ago to pantomime tooth brushing is attributed to something amiss in the "brushing the teeth" program. Such an explanation has a lot to recommend it. Much of what we do depends upon the establishment of a routine, i.e., motor program. Tying our shoes and our neckties involves related but different routines that must be laboriously learned step by step. After learning them, we no longer have to concentrate on the individual movements involved (in fact doing so usually interrupts the sequence and we must begin again). Apraxia thus might represent one or more failures in the smooth operation of the learned routine. But despite their initial appeal, such explanations fail to account for why a sequence that fails in the clinic later occurs normally in a more natural setting. In addition, even the simplest of our acts is very rarely, if ever, repeated in precisely the same way. My reaching out to pick up the telephone now ringing on my desk varies according to my distance from the phone, the positions of my eyes and hands, what I am doing at the moment, even whether or not I care to be interrupted—which, if we want to talk in terms of motor programs, might be thought of as activating an "ignoring the telephone" program *inhibiting* the muscles employed in reaching for the phone. The "telephone answering program" must accommodate the recruitment of different muscles and muscular interactions each time the phone is answered. The "program" must be based on some representation within the brain of the initial state (the phone sitting on my desk) and the final state (the receiver held up to my ear). Moreover, these "programs" must remain fluid and dynamic.

Dynamic patterns are involved in the simplest movement, and these patterns reflect the complicated interplay of vectors of force and motion expressed over time by shifting patterns of muscle groups. Thus the word "program" is perhaps an unfortunate one, since the word as usually employed refers to a set and comparatively invariant sequence (as in "We attended a Bach program at the Kennedy Center"). Certainly the "program" (we seem stuck with this term which is firmly encrusted in the writings and utterances of brain experts) must be an extremely fluid one that is based more on *intention* than on the activation of specific muscles. And in order to move from one state to another, updates must be constructed from moment to moment, millisecond to millisecond. The location at any given instant of the phone receiver on its trajectory toward the ear is signaled to the brain by means of messages carried by sensory receptors in the fingers, wrist, shoulder, and other locations. Somehow my desire to pick up the phone activates a loosely defined "program" which selects from an almost infinite number of possible patterns of muscular excitation and inhibition. Handwriting is another example. Writing my name on a paper while sitting at a desk results in a signature identical except for size to the one that results when I stand at a blackboard and write with chalk. Yet each of these identical signatures involves different programs composed of the actions of different muscles (the muscles of the shoulders and forearms aren't used at all when I'm writing at a desk). But to speak of "programs" presupposes a "programmer," and with it the return in a more contemporary guise of the dreaded homunculus: Who or what is writing the "program" and where in the brain is this taking place?

After a goodly number of false starts, many neuroscientists believe we are now in a position to answer that question. They relegate the homunculus to the old model of information processing where, like our liquor-pouring example mentioned above, everything occurred sequentially. In such a model, somebody or someone has to start the process off, hence the ever-present homunculus. But if the true state of affairs involves a wide network involving many mutually interactive processes occurring at the same time (distributed processing, as it's referred to in information transfer terms), then the conundrum of the homunculus is solved. Or so it's claimed. In order to make up our own mind about this, consider some additional facts about what happens in our brain when we make our "simple" movement of reaching to pick up our ringing telephone.

Human reaction time tests provided the basis for our most popular view about the control of motor behavior. Subjects respond to a sensory signal, a sound or flash of light, by making some kind of movement, usually pushing a button. Their responses are divided into a reaction time—the period from the stimulus to the movement, and a movement time—from the onset of the stimulus to the completion of the movement. Since the reaction times are longer than synaptic delays (the time needed to send a signal from one neuron to another), some explanation is called for. Traditionally it is that the brain goes through a time-consuming sequence of operations: detect and analyze the stimulus; decide which movement is called for; program the movement; finally, send the programmed command to the appropriate muscles. With the onset of movement, the reaction time ends. According to this model, information transfer involves the enactment of a sequential series of operations somewhat along the lines of the simplified explanation I presented above. But recent brain research indicates that things proceed quite differently.

Information transfer within the brain is not like a train moving from station to station but more like the spread of a rumor. Transfer takes place simultaneously in multiple brain areas via a widely distributed network of nerve fibers.

Electrical recordings from motor and premotor neurons in awake animals reveals that a very large fraction of motor cells respond during any arm movement. Not all of them are totally activated ("fire"), but some degree of response occurs, with the final arm movement related to the sum of activities of the entire population of motor cortical neurons. Thus the motor command to move is spatially distributed throughout the cortical population. Not only that, motor cells are responsive to sensory stimulation, with each cell biased toward light or sound or other sensory stimulus. During the movement, the sensory and the motor components are so intertwined that the old division into sensory and motor neurons no longer makes sense. Limb movements result from the activation of these sensory-motor combinations which interact with each other through feedback loops. In this way the slightest of sensory signals can initiate activity in a vast network of neurons. Moreover, different neural systems are involved in initiating, programming, and terminating a particular movement. Nor is this process limited to the cerebral hemispheres. Extensive two-way communications exist between the hemispheres and subcortical structures. As a result, no "command" neuron or network of neurons is responsible for the movement, which unfolds with the inevitability and internal logic of the opening of a flower.

To summarize: If free will is what differentiates us from the rest of the animal kingdom, an assertion which I am prepared to accept, then it would certainly be nice to be able to say something along the lines of "And *this* part of the brain tunes up when we act freely." But observations of patients suffering from disorders of the will, along with electrical recordings from perfectly normal people engaged in willed action, fail to support the notion of a "will center." Difficulties arise, it seems, whenever we attempt to pin down either the location or the timing of our most deliberated actions. What's more, if we suffer some form of apraxia, we may be unable to stop ourselves from doing and not doing some of the strange things I described earlier. Human will, if I can be forgiven the use of such a pretentious term, seems to operate in a modular fashion and is composed of any number of submodules. I am now thinking in particular of Kenneth Heilman's patients, who cannot respond to spoken requests ("Show me how you would use a hammer."), and yet when they are simply shown the instruments, everything goes off without a hitch. Or those who incorrectly sequence the cleaning of a pipe so that they might first light the pipe, then put the tobacco in, and then start cleaning it.

With such disturbances we enter into the existential realm that philosophers are so keen on discussing. Just consider what it must be like to live in a world in which you cannot decide on something, set about doing it, and proceed in the correct order. One difficulty in carrying out such an exercise of the imagination and putting oneself in that person's place is that few patients can tell us about their inward states, since language disturbances usually accompany the apraxic disorders.

Our reliance on language explains some of the difficulty we have in coming to terms with apraxia or the research findings of Kornhuber or Libet. The concept of ourself as a unified, freely acting agent directing our behavior is firmly entrenched in our written and spoken language. But even a casual effort at introspection reveals that even the most balanced of us are often of two or more "minds." One part of us wants desperately to do something, while another part resists with a ferocity that leaves us feeling disjointed and conflicted. At such times we wonder if more than one person occupies our bodies. Brain research on consciousness carried out over the past two decades casts important doubts on our traditional ideas about the unity and indissolubility of our mental lives.

CHAPTER TEN

The Flip of a Coin

Consciousness resists definition partly because it is so familiar. We experience it ourselves and by analogy infer it in others. Nor can it be defined separately from perception, learning, thought, memory, and emotion. At all times we are conscious *of* something (the word derives from the Latin *cum* (with) and *scire* (to know). In response to the difficulties defining consciousness, some have opted to dispense with it all together.

"Literally hundreds of thousands of printed pages have been published on the minute analysis of this intangible something called 'consciousness.' This thing we call consciousness can be analyzed only by introspection—a looking in on what takes place inside of us. As a result of this major assumption that there is such a thing as consciousness and that we can analyze it by introspection, we find as many analyses as there are individual psychologists," wrote James Watson, who expelled consciousness from his behaviorist psychology.

Watson had a point: that one person's consciousness is different from that of others. But this does not prove that consciousness doesn't exist, only that we have difficulty representing it to ourselves and describing it to others. As Saint Augustine pointed out prior to Descartes, you have only to doubt your own consciousness to have established it. Nonetheless explaining one's consciousness to others is quite a different matter. In his *Philosophical Investigations*, Wittgenstein compared consciousness to "a beetle in a box." "No one can look into anyone else's box, and everyone says he knows what a beetle is by looking at *his* beetle. . . . Everyone might have something different in his box. . . . The box might even be empty."

But despite the difficulties, consciousness is not a total muddle. Like the beam of a searchlight, consciousness consists of both central and more marginal components. And what is central and what is marginal may change from moment to moment according to the situation. A creaking on the stairwell may dominate our consciousness in the middle of the night, but that same sound heard during the day hardly attracts our attention—a prerequisite for consciousness. At all times the separation of background from foreground in consciousness changes in relation to our verbal capacity to describe inner experience (a talent that, as we shall see, differs considerably from person to person). In any case, you can only describe your consciousness by a process of selection. Certain aspects are always emphasized more than others. This is the natural consequence of the intentional nature of consciousness.

While William James referred to consciousness as "the faint echo left behind by the disappearing soul," Dostoyevsky, in his partly autobiographical novel, *Notes From Underground*, warns "I am firmly convinced that a great deal of consciousness, every sort of consciousness, in fact, is a disease. . . . To be too conscious is an illness—a real thoroughgoing illness." Perhaps the safest thing to say on the subject is that consciousness, like other mental states, is best in moderation. "Consciousness-expanding" groups and books notwithstanding, consciousness, particularly when it merges into morbid and unbalanced *self*-consciousness, can be very akin to an illness.

So far no one has set forth a convincing argument that consciousness, however it is defined, has ever existed in the absence of a brain. It's true that philosophers and psychologists have argued over the centuries about the relationship of the mind and the body (now almost universally acknowledged as involving the mind and the brain), but with the exception of pure Idealists like Bishop Berkeley, none has dispensed with the brain altogether. Recently some philosophers have claimed that consciousness would emerge with the achievement of a given level of complexity in a nonliving system. We will say more later why this is unlikely, but for now it's sufficient for my point to note that according to everything encountered until now, consciousness requires a brain. While electric doors, thermostats, calculators, and computers exhibit an "awareness" of various aspects of the environment, no serious claim can be made, except by analogy, that, for instance, a thermostat turns back a furnace because it is "conscious" of feeling overheated.

If consciousness is related to the brain, what part of that organ is most important? Traditionally, it was generally accepted that consciousness was bound to the cerebral cortex. This inference followed from the reasoning that since consciousness was humankind's most advanced and sophisticated mental attribute, it must emanate from the most evolved portion of the brain. Insight into the true state of affairs started with some observations made during the 1918 worldwide influenza epidemic.

Patients with a special form of brain inflammation (Von Economo's encephalitis) exhibited extreme and bizarre disorders of consciousness. They spent most of their time asleep and had to be aroused for meals, after which they immediately fell back again into a stuporous state. Not only did many patients lose consciousness along a continuum from extreme stupor to profound coma, but many of them even when awake lost all sense of identity and separation from the people and things around them. "What is what, did you say did I say, what word is that word what, what do words the word what mean?" was one exclamation by an encephalitis patient (his doctor incidentally characterized this strange utterance as "so much reminiscent of the literary output of Gertrude Stein"). At autopsy, inflammatory changes of encephalitis did not appear in the cerebral hemispheres but in the midbrain, an underlying structure in the upper brain stem.

Two decades later a Belgian physiologist working with cats severed the connection between the cerebral hemispheres and the lower brain stem. Absent the influence of the cerebral hemispheres, the cats, although almost totally paralyzed, remained awake and their electroencephalograms (EEG; a measure of electrical brain activity) showed normal sleeping and waking activity. This established that wakefulness, a necessary prerequisite for consciousness, does not depend upon the cerebral hemispheres.

Definitive proof that the cerebral hemispheres alone are not the mediators of consciousness came in 1949 in an experiment with an anesthetized cat. The experimenters, Giuseppe Morruzi and Horace Magoun, stimulated an electrode sitting in the reticular formation, a complex network of nerve fibers and cell bodies in the core of the brain stem. At that moment the cat woke up and its EEG shifted from a sleep to a waking pattern. Destruction of the reticular formation resulted in a permanently comatose animal. Additional research revealed a system of fibers linking the reticular formation with the thalamus, hypothalamus, and related structures. This reticular activating system serves to maintain arousal and thus is concerned primarily with the maintenance and regulation of consciousness, while the cerebral cortex is more closely connected with the contents of consciousness. But this division of labor is highly artificial, since consciousness cannot exist in the absence of wakefulness. Nor can it long endure in a normal state under conditions of sensory isolation. Laboratory experiments as well as natural disasters and mishaps resulting in prolonged sensory isolation regularly induce distortions of consciousness (hallucinations, illusions, etc.) Again: Consciousness must always have a content and a relationship to that content. One may be conscious of the environment, one's own body, emotional state, intentions, past, present, and future, even one's momentary self-experience, "the consciousness of one's own consciousness"—and this is just a partial listing. All of these examples involve the interrelated activity of the reticular activating system and the cerebral cortex. But in no instance can consciousness be said to emanate from a center or special area within the brain. Indeed consciousness is not a "function" that lends itself to localization but, rather, an emergent property of neurons at a certain level of complexity.

"What counts for subjective unity may lie in the way the brain processes function as a unity regardless of the multilevel or multicomponent make-up of the neural events involved," according to Nobel Prize–winning neurobiologist Roger Sperry, who is best known for his work on the effects of "split brain" operations. The operation that separates the right and left hemispheres from each other has inspired a whole mythology. Artists and architects are described as "right brain" types, while writers and others who employ language find themselves described in terms of "left brain" superiority. Most of this is nonsense, of course, since in Sperry's words, "It is important to remember that the two hemispheres in the normal intact brain tend regularly to function closely together as a unit and that different states of mind are apt to involve different hierarchical and organizational levels or front-back and other differentiations as well as differences in laterality." But I wish to discuss split brain or, more correctly, post-commissurotomy patients, in order to make a different point. While wakefulness, awareness, attention, and primitive forms of consciousness may result from the action of many parts of the brain working together, the more sophisticated states of consciousness seem to depend on the cerebral hemispheres. Let's briefly review what twenty years of split brain operations have taught us about consciousness.

If I introduced you to a person who has undergone such an operation, it's doubtful that you would notice anything unusual about him. Despite having undergone severance of the largest fiber system in the brain— some 800 million fibers—his appearance, speech, and everyday behavior are completely normal. Upon recovery from this operation, most patients are able to return to school or to an undemanding job. Two years after surgery, the person could easily pass a general physical exam and score normally on the verbal portions of IQ tests. To this extent he or she appears to be the "same" person as before the operation.

"The general behavior and conversation during the course of a casual encounter without specific tests typically reveals nothing to suggest that these people are not essentially the same persons that they were before the surgery and with the same inner selves and personalities," according to Sperry.

Because the patients looked, talked, and acted normally after their surgery, early investigators concluded that they *were* normal. But more careful analysis aimed at evaluating the actions of each of the two separated hemispheres revealed profound differences. The right hemisphere is dominant for nonverbal, largely spatial tasks like copying designs, interpreting facial expression, mentally transforming or transferring visual images in one's mind, intuitively appreciating geometrical designs. In addition, the right hemisphere is superior to its counterpart in expressing and appreciating emotions. The left hemisphere is the speech and language maven: it handles reading, writing, and understanding spoken language. The left hemisphere is also specialized for calculation. But more important than the parceling of function is the fact that each hemisphere has its own unique consciousness. In the case of the left hemisphere, consciousness is linked with language capacity. If you ask questions, it's the left hemisphere that understands the question and replies. But the right hemisphere's abilities (the intuitive apprehension of geometrical properties, copying designs, recognizing faces, and reading facial expressions) clearly imply that some degree of consciousness, albeit a nonverbal one, must exist.

In a typical experiment testing the separate abilities of the two hemispheres, a split brain subject is shown what neuroscientists call a *chimeric picture*: the halves of two different photos joined together in the midline to make a double picture. One chimeric picture, for example, shows the right face of the actress Catherine Deneuve joined to the left face of an elderly Italian worker from Fisherman's Wharf in San Francisco. When the subject looks at the chimeric picture, the actress is seen by the right hemisphere while the bushy-browed fisherman is seen by the left hemisphere. The subject is asked to identify what he has just seen. The answer depends on how the inquiry is conducted. If the subject is asked, "Point to what you saw," the subject chooses the picture of the actress. This is because the right hemisphere is utilized in tests of facial recognition. But if, instead, the subject is asked "What did you see?" he or she answers "A man with dark hair and heavy bushy eyebrows." Words, sentences, and questions bring the left hemisphere into play.

Neuroscientists turned up an additional remarkable insight into the brain from the chimeric picture experiments. Despite the fact that each hemisphere of the subject saw only one side of a face, it imaginatively transformed the half face into a whole. When asked what they had seen, the subjects never replied, "One half of the face of a man with dark hair . . . , etc.," but always a *complete* face. And those who pointed to a picture of Catherine Deneuve never indicated in any way that they had seen only one half of her face. The brain here is performing a kind of "filling-in process," that enables it to guess a complete picture from only partial information. A similar filling-in happens every time we look at anything. None of us is aware—except during special testing procedures—that we are blind in each eye at a point about 15 degrees lateral to our center of gaze. This point corresponds to the exiting of the optic nerve from the eye on its path to the brain. Since there is no room for receptors in this area, it is a "blind spot"; but we don't experience any lack of vision, since the brain completes the scene for us according to its best "guesstimate."

Chimeric pictures and blind spots illustrate that "reality" is at least partially constructed for us by the brain's inherent organizing principles. None of these involves consciousness or awareness; indeed, we have to construct special and highly artificial test situations in order to discover them. The Gestalt psychologists in the 1920s and 1930s formulated "laws of organization" that seem to conform to the findings of chimeric pictures and the blind spot.

"After watching repeatedly the superior performance of the right hemisphere in tests of design one finds it most difficult to think of this half of the brain as being only an automaton lacking in conscious awareness. It is especially difficult to deny consciousness to the right hemisphere where it proves to be superior in novel tasks that involve logical reasoning and also when it generates typical facial expressions of satisfaction at tasks well done or of annoyance at its own errors," says Sperry. He feels that the right hemisphere, despite its underdeveloped language capacities, harbors a well-developed, seemingly normal consciousness, basic personality, and social self-awareness.

Sperry's work on split brain patients prepared the way for the recognition that consciousness can exist independent of awareness. At first such a distinction seems to make no sense at all. In Chapter Four we encountered people like Helen French who were selectively unaware of things around them, even of the extent of their injury. But the split brain patient takes things a bit further: The right hemisphere is conscious of important distinctions along the lines of geometry, form, and the appreciation of emotional expression—so the performance of the person indicates—yet if asked about them, consciousness is denied. We encounter here a split of consciousness, in which something is both affirmed and denied. The right brain makes the correct decisions, but the person cannot consciously explain how that was done. "For the commissurotomy subject direct awareness is no longer whole," according to Colwyn Trevarthen, who worked with Sperry in some of the original research on split brain subjects.

Freud spoke of processes that influence behavior and remain permanently outside of consciousness, which he termed the *unconscious*. But his contribution suffered a credibility problem when he insisted that the unconscious is somewhat like a boiling cauldron of inhibited and repressed primal instincts such as aggression and sex. But this isn't true. Split brain patients remain "unconscious" of all kinds of things that have nothing to do with sex or violence. Neuroscientists speak of the "cognitive unconscious," and we have encountered it in various forms throughout our exploration of the modular brain (anosognosia, implicit memory, to name just two). But the most interesting example of split consciousness involves retained "unconscious" visual abilities that operate outside of their conscious awareness in the permanently blind.

If a blind person is asked in a test situation to "guess" at the location of a light flashed on the wall by an experimenter, he will guess correctly 80 percent of the time. Most fascinating, he will deny seeing anything in his blind visual fields, though if pressed for an explanation, he will speak of a "feeling" about the location of the light. Lawrence Weiskrantz, the researcher who discovered "blindsight" (as he refers to the phenomenon), attributes it to the action of one of the other nine branches of the optic nerve that connect with regions of the brain other than the visual cortex. The most likely candidate is a large branch which goes from the eye to a way-station in the midbrain called the *superior colliculus*. In people with normal vision, this tiny, rounded pro-tuberance activates outside of awareness whenever light is projected anywhere in the visual field. With evolution of the primate brain, the path from retina to superior colliculus receded in importance. A monkey blinded by a surgical excision of his visual cortex will learn to see again thanks to the "backup system" provided by the superior colliculus. But if that same monkey is reoperated on so that the superior colliculis is destroyed, he will remain permanently blind. Weiskrantz and other researchers believe "blindsight" results from activation of the superior colliculus.

Although the superior colliculus is less important in the human brain than the monkey brain, when the visual centers in the cortex are dam-aged it can exert a powerful influence on vision, if that damage occurs early in life. An eleven-year-old boy who had his occipital lobe surgical-ly removed because of a tumor retained near normal vision after the surgery. Moreover, he retained normal visual awareness.

Obviously a distinction must be made here between awareness and consciousness. While consciousness implies awareness, the relationship is not reciprocal. We can respond to something, implying some level of awareness, yet we may remain blithely unconscious of what's happen-ing. In a movie theater, for instance, the dimming of the lights usually elicits a vague and then clearer "consciousness" of diminished bright-ness. But this distinction is often artificial, as with one of Weiskrantz's blind patients who felt as if something was approaching him when a light went on in his blind visual field and felt that something was reced-ing when the light went off. At other times he had no feelings at all and yet correctly guessed the timing and direction of the light.

Disorders corresponding to "blindsight" occur, although rarely involving other sensory channels. A patient suffering a stroke which deprives her of awareness of touch on one side may, when blindfolded, successfully point to an area on the anesthetic skin where she has been touched. Consciously she doesn't feel a thing, yet her performance clearly implies awareness. Similar dissociations occur in hypnosis. A hypnotized person may deny feeling pain when stuck with a needle, but measurements of pulse, blood pressure, or other physiological indices show that the stimulus has nonetheless been perceptually registered. And careful testing of subjects in post-hypnotic amnesia (the person denies memory for events transpiring during the hypnotic state) reveals that the relevant information entered the brain and is accessible for retrieval, although not necessarily *conscious* retrieval. Even post-hypnotic suggestions (the subject later carries out some action suggested by the hypnotist) interfere and compete with post-hypnosis mental activities—a sure sign of subliminal awareness.

"Since post-hypnotic suggestion takes place outside of phenomenal awareness, it appears there are circumstances in which complex, deliberate, attention-consuming processes may operate nonconsciously," according to John Kihlstrom, professor in the department of psychology, University of Arizona, who has made a special study of what he refers to as the "cognitive unconscious."

After a career devoted to consciousness, Kihlstrom has concluded that "Consciousness is not to be identified with any particular cognitive-perceptual functions, such as discriminative response to stimulation, perception, memory, or thought. Rather, consciousness is an experiential quality that may, but need not, accompany even complex information-processing activities. Nor is consciousness to be identified with focal attention . . . but seems to require that a link be forged between an activated mental representation of an event and an activated mental representation of oneself as the agent or experiencer of that event."

Kihlstrom emphasizes that consciousness requires a vivid awareness of oneself as an *experiencer*. The absence of such awareness accounts, I would suggest, for many of the more dramatic instances of multiple personality, and the so-called "fugue states." In this state a person wanders away from his usual surroundings, and several days, weeks, or even months later suddenly "comes to" in a strange place and without any clue as to how he got there. I've encountered several patients with fugue states in my neuropsychiatric practice, and all of them were escaping from difficult experience they could not master. When their coping abilities failed them, they escaped from the unbearable situation by blocking off their awareness of themselves as experiencer and initiator. I don't mean to imply they did this deliberately. Rather, certain individuals, for reasons we don't understand, are prone to deal with extreme stress by *dissociating* aspects of their mental lives. To this extent, they are carrying to an extreme what we all do at times: go off on a trip or seek new surroundings in order to "forget about" some problem.

Hypnotic subjects and their dissociative abilities bring home to us the importance of subjectivity. And this creates a problem, since as a culture we have a love-hate relationship with subjectivity. Politically and culturally we prize it; our First Amendment guarantees everyone's right to hold and express highly subjective opinions; imagine the entertainment industry if subjective interpretations were abandoned and our movies and television shows were created for us by computers. At the same time and in other areas, we place severe limitations on subjectivity in favor of objective measures of competency (such as licensing requirements for professionals) and truth and accuracy (such as our legal system, or the Food and Drug Administration's long and detailed investigations into whether a drug is safe and effective before it is allowed on the market).

Our ambivalence about subjectivity can be traced to Galileo, who asserted in 1623 that science should be concerned only with qualities that can be weighed and measured. What he described as "secondary qualities"—value, meaning, love, and beauty—fall outside the concerns of science, he maintained, and should be ignored. Descartes picked up on this theme and divided the world into two starkly different kinds of substance. The first extends into space (*res extensa*) and, since it has dimensions of length, breadth, and width, can be measured. The second, which he referred to as the "thinking substance" (*res cogitans*), has no physical characteristics and so can neither be measured nor divided. All of reality, of which the human brain is one component, belongs to the first division, the mind and consciousness to the second.

From Galileo and Descartes we inherited our present-day distinctions between mind and matter, subjective and objective. And since until recently the subjective operations of consciousness could not be measured, the reality of consciousness was either denied altogether (the behaviorists) or treated as an awkward inconvenience. But, as I hope I have convinced you by now, subjective processes *can* be measured. Recall the experiment described in Chapter Three where the *intention* to move the finger produced PET scan activation of the finger area of the motor cortex, or the experiments concerning willed action carried out by Libet (see Chapter Nine). As a result of experiments like these, subjectivity has become respectable again. We make a big mistake, however, if we assume that subjectivity can be completely "objectified" so that, for instance, thoughts may be understood in terms of bioelectrical patterns or neurotransmitter interactions. In my previous book, *Receptors*, I discussed in considerable detail why "crooked" thoughts are unlikely ever to be understood strictly in terms of "crooked" molecules. But my purpose here is to plumb a bit deeper and show how we continue to underestimate the importance of subjectivity. In a phrase, the only world that we know, the only world that we *can* know is a subjective one.

As I look up from writing this passage and stare out into the woods across the street, I experience a world of trees, sunlight, and the sounds of birds—sensory objects, as a philosopher would term them. Yet what I actually experience is energy in the form of vibrations from waves of different frequencies: low frequencies for hearing, higher frequencies for warmth, still higher for vision. At this level of "reality," subjectivity does not exist.

As these waves interact with the receptors in my ears and eyes, they trigger neural codes which upon relay to the brain result in the brain's creation of a model of the external world. Since it is the brain that does this, our individual brains, the model assumes a subjective character: It is *our* world. Yet we don't think or speak about it as subjective; on the contrary, this mixture of the objective (waves of energy) and the subjective (our brain's construction via neural codes of a model), we refer to as the "objective world." But it is not objective in the sense of being known independently of the brain. Recall the example of that tree falling in the forest. It makes no sound but produces only wave forms which, when striking the tympanic membrane, set off the neural code our brain interprets for us as sound. What is the nature, the "reality" of that wave if no ear is present to transmit it into a neural code? There is no answer to that question because those conditions cannot be accommodated in the world modeled for us by our brain.

When it comes to appreciating the subjective nature of so-called "objective" events, physics is about sixty years ahead of neuroscience. According to the Heisenberg principle of uncertainty, any attempt to measure the location of a subatomic particle, such as an electron, disturbs the particle so much that its momentum cannot be determined. Conversely, measurement of the momentum results in loss of information about the location of the particle. If measurements of momentum and location are taken simultaneously, a compromise is required: foregoing momentum knowledge to gain position knowledge, or vice versa. In these experiments, subjectivity and objectivity interpenetrate; we can only appreciate "objective" qualities by measuring them with "objective" instruments whose measurements require some type of "subjective" interpretation. When we look through microscopes at brain cells or examine chemicals withdrawn from a lobe of the brain, we are not making an observation of an "objective" world but of a subjective perceptual model created for us by our brain. No matter how many measuring instruments may be involved (perhaps a computer is carrying out all the measuring), at some point all of this information must be "understood" in terms intelligible to the human brain; perhaps only a handful of brains can manage the task of understanding some of Einstein's work. But however many brains are capable of "understanding" objective scientific observations, the ultimate operation is a subjective one, involving neural codes and models constructed by the brain.

Since few of us have backgrounds in quantum physics, let me mention an example of the power of subjectivity more likely to come up in everyday life. The flip of a coin involves, according to probability theory, an "objective" science par excellence: a 50 percent probability that the coin will be heads or tails. After a specific flip, the probability of heads is 50 percent. But suppose that at the moment the coin lands, an observer sees that it is up heads. What then is the probability? In truth, there are now two probabilities instead of just one. In my case, not having seen the coin, the probability is still 50 percent. The same probability would hold true for every other person in the world, *save one*. For the observer, and directly as a result of his observation, the probability is 100 percent for heads, zero for tails. In this example, an aspect of "objective" reality, probability, is modified subjectively. Knowledge renders probability considerations useless.

My point here is that subjectivity is not some second-best means of encountering the world, forever overpowered by an "objective reality" which is to be measured and yielded to. Rather, any experience that arises in consciousness as a subjective experience, is endowed with its own reality, and can exist independent of, and even, as with the coin-flipping example, in conflict with, "objective" reality. Put another way, consciousness must be understood as a very special *emergent* property of the human brain. It is not an indispensable quality, since as we have seen the vast majority of the brain's activities do not involve consciousness. It is not always a desirable property, at least in its more self-referential forms (conscious of myself as consciously thinking about consciousness, and so on in an infinite regress, as with some of the characters depicted by Kafka and Dostoyevsky). But consciousness is a unique property of the brain made possible by a sufficient number of parallel interacting modules. At this point, we stub our toe on imponderables: How many modules? how arranged? where? interacting according to what principle? No brain scientist presently living has the answers. Will anyone ever come up with these answers? One hesitates to say "Never," of course. I'm thinking again of that man unreeling and examining the threads of twine that, unknown to him, constitutes the matter comprising his own head. And there is also another factor of importance that, so far, we haven't said much about. We are not just creatures who think, we also *feel*, experience, and express emotions about things. On most occasions we are not only conscious of things, but conscious of how we feel about them as well. At this point I propose we step back from these reflections on consciousness to explore the possibilities and limitations on our brains imposed by those modules concerned with emotion.

CHAPTER ELEVEN

The Papez Circuit

Affects differ from other forms of psychic information insofar as they are subjectively qualified in a physical sense as being agreeable or disagreeable . . . There are no neutral affects because, emotionally speaking, it is impossible to feel unemotionally.

Brain researcher Paul MacLean

Our internal feelings or affects are private and subjective, while their expression, our emotions, are on occasion all too public and obvious to others. But most of us develop at least some rudimentary skills in keeping our feelings well hidden and, most of the time, can prevent their overt expression. Certain of us become highly skilled at this, with the most advanced practitioners of the art even capable of hiding feelings from themselves. But if our deception of others thus evolves into self-deception, both our feelings and our emotions disappear altogether from our mental landscape. At such times we establish for ourselves a bleak inner world, cease to experience feelings, lose the capacity to express emotions, and encounter first hand the paradox that MacLean points out: Internal feelings can never be neutral, and one cannot feel unemotionally.

Our ideas about emotion can be traced back at least as far as the Greeks. Plato described the inspiration that flows from the poet to the actor and then to the audience as like the power of a magnet. But modern thoughts about the emotions started with our old friend Descartes. To him, the emotions were part of the immaterial mind. But in place of "emotions," he spoke of the "passions" of the soul (mind). He believed the mind is a nonphysical substance linked with the brain via the centrally placed pineal gland. Although Descartes remained vague about how a physical organ can link with a nonmaterial substance, his theory of the origin of the emotions, set forth in *The Passions of the Soul*, provided the first explanation of emotions based on the brain.

"Let us then conceive here that the soul has its principal seat in the little gland which exists in the middle of the brain, from which it radiates forth through all the remainder of the body by means of the animal spirits, nerves, and even the blood, which, participating in the impressions of the spirits, can carry them by the arteries into all the members."

Descartes set out his theory of emotions in response to the question whether a mental state of fear arises between seeing a wolf and fleeing from it. He asserted that the brain responds automatically to the danger signal carried to the brain along the optic nerves. Upon receiving the signal, the brain drives "vital spirits" downward along the descending nerves to produce the muscular action of running. The brain is then informed of the flight by means of sensory fibers from the muscles and nerves in the legs. This information is conveyed to the pineal gland which communicates directly with the soul, and emotion (passion) is experienced. In this model, the emotions are the mental consequences rather than the causes of the muscular and glandular actions involved in the flight from the wolf.

Descartes's model of the passions would be thoroughly neurological but for one thing: To him, the mind was separate from the brain. While the brain was part of the body, the mind was a completely nonphysical entity. (In Chapter One we discussed the logical *cul de sac* Descartes created for himself with this tenet, but that is not our subject now.) This split between body (brain) and mind entailed certain consequences. Passions (emotions) resulted from influence of the body on the mind by means of the brain's pineal gland. And, Descartes insisted, influences could proceed in the opposite direction: the mind influencing bodily action through the action of the intervening pineal. Such influences could come about, however, only as a consequence of reason and never of passion. This was the origin of the split that remains today between reason and emotion. To Descartes and his followers, the Cartesians, humankind's greatest challenge is to bring the passions under control by means of reason mediated by the incorporeal soul. Over the ensuing centuries, thought and emotion have occupied separate domains. Wise men of all cultures are typically portrayed as "reasonable," "rational," and governed by the dictates of reason. To the extent that a person deviates from this ideal, he or she has been judged less than human.

Two hundred years after Descartes, his theory of emotions (we do not run from the wolf because we are afraid but we are afraid because we run) appeared anew in the ideas of the American psychologist-philosopher William James and the Danish physiologist Carl Lange. By then knowledge about the brain had progressed dramatically. It was known that motor and sensory nerves travel separately to the brain and terminate in different parts of the cerebral cortex. Experimental studies with animals had established pathways for sensation and muscular action which are still valid today. But as William James noted, "the aesthetic sphere of the mind, its longings, its pleasures and pains, and its emotions, have been so neglected in all these searches."

James concluded that "aesthetic" processes result from the operation of the sensory centers. Perception of a stimulus leads to some muscular activity, and our awareness of these changes as they are occurring *is* the emotion. Lange's version of the origin of emotion emphasized a similar sequence:

> If I start to tremble when I am threatened with a loaded pistol, does a purely mental process arise, fear, which is what causes my trembling, palpitation of the heart, and confusion; or are these bodily phenomena aroused immediately by the frightening cause, so that the emotion consists exclusively of these functional disturbances of the body?

Note that both James and Lange rejected the idea of specific emotional centers within the brain. In their view, emotion could be adequately explained as just another function of the sensory and motor areas somehow working together. Despite the tidiness of such a theory, evidence was available to James that contradicted it.

In 1715 the Dutch physician and chemist Herman Boerhaave had observed that even perfectly reasonable people, when bitten by rabid animals, began "gnashing their teeth and snarling like a dog." At autopsy the victim's brains were swollen and markedly inflamed. Two centuries later Boerhaave's discovery was clarified when a French neurologist demonstrated that rabies, by then known to be caused by a virus, infects areas beneath the cerebral hemispheres. Additional evidence was provided by Sir Charles Sherrington, the British neurophysiologist best known for his poetical characterization of the brain as "an enchanted loom." He showed in a series of experiments—usually described as "elegant," but which were also pretty grisly—that a cat with the top of its brain cut off still displayed emotional reactions when aroused, clear proof that emotional expression isn't dependent on a cerebral cortex. As long as portions of the thalamus and hypothalamus remained intact, emotional reactions could be elicited from the animals. A similar situation was described in humans by the British neurologist Sir Henry Head. He observed that patients who had suffered strokes in the thalamus on one side reacted excessively to very slight stimulation. Thus a mildly disagreeable stimulus such as a pinprick evoked in the patient extreme discomfort. Head concluded that the thalamic lesions had released that structure from cortical inhibition (recall from Chapter One John Hughlings Jackson's tenet that the cerebral cortex inhibits rather than stimulates the lower centers).

Following on the heels of Sherrington's and Head's observations, neurophysiologists Walter Cannon and Philip Bard argued for a special emotional center within the brain and shifted the emphasis from the thalamus to the hypothalamus. They suggested that the hypothalamus encoded the emotional components of the sensory information transmitted from the thalamus. From the hypothalamus, information then went upward to the cortex to evoke emotional experiences and downward to the brain stem to induce emotional behavior. Experimental proof for the Cannon-Bard theory of emotion soon followed.

Electrical stimulation of the hypothalamus of cats carried out by Walter Hess of Switzerland turned tame pussycats into enraged Halloween cats that responded with hissing, scratching, fighting, and other attack and defense maneuvers. Once the hypothalamus itself was destroyed, only components of the rage reaction could be elicited. Snarling, spitting, tail lashing, and increases in heart rate and blood pressure occurred, but the animal no longer attacked. This abbreviated response was referred to as "sham rage," since the rage behavior was no longer part of a fully integrated pattern of directed attack.

Hess's detractors initially raised doubts that electrical stimulation elicited natural states of rage. But Hess insisted that a center for genuine anger exists in the hypothalamus, at least in cats. "There is no reason to consider it different from natural rage," he wrote several years before receiving from his peers the ultimate vote of confidence in his work, the 1949 Nobel prize in physiology.

During this same period, an American neuroanatomist was solving the mystery I alluded to a moment ago: how a perfectly normal person, as the result of being bitten by a rabid animal, could be transformed into a snarling beast.

"The prodromal symptoms—insomnia, irritability, and restlessness—usher in the stage of excitement and profound emotional perturbation. . . . Light and sound, and every stimulus situation provokes great apprehensiveness and paroxysms of fear. The patient presents the appearance of intense fright and of mingled terror and rage," wrote J. W. Papez. By studying the brains of rabies victims, Papez observed that the lesions remained confined to certain subcortical areas. This offered, in Papez's words, "an important clue to the probable location of the emotive mechanism."

In 1937 Papez proposed as the brain's mechanism of emotion a circuit centered around the hypothalamus. The Papez circuit, as it came to be called, is composed of a collection of interrelated structures that include the *hippocampus*, the *fornix* leading to the *hypothalamus*, the *mammilthalamic tract* leading to the *anterior thalamic nuclei*, and the *cingulum bundle* leading to the *cingulate gyrus*. Papez's purpose was to explain how sensory information that passes through the thalamus en route to the cerebral cortex takes on emotional significance. He suggested three exit pathways from the thalamus: to the cortex, responsible for the "stream of thought"; to the basal ganglia below the cortex, responsible for the "stream of movement"; and to the hypothalamus, responsible for the "stream of feeling." He agreed with Cannon and Bard that impulses from the hypothalamus were transmitted upward to the cortex or downward to the brain stem and spinal cord where they could evoke autonomic emotional reactions even, as with the animals surgically deprived of their upper brain areas, in the absence of a cerebral cortex.

According to Papez, the projections upward from the hypothalamus and by way of the anterior thalamic nucleus to the cingulate cortex provided the emotional underpinnings of "psychic" processes occurring in the cortex. Because these brain areas form an interconnecting circuit the hypothalamus could be activated either by direct sensory stimulation from the outside world via the thalamus or the emotion could originate in the cerebral cortex in the form of an emotion producing thought and work downward. Thus Papez's circuit relegated emotional experience to

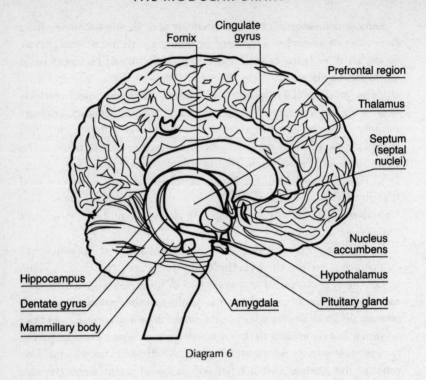

Diagram 6

the cerebral cortex and emotional expression to the hypothalamus and its subcortical connections.

A decade later, additional structures were identified by Paul MacLean as important in emotion. He spoke of a functional system within the brain important in emotional feeling and expression. MacLean called it the *limbic system*. A ringlike structure on the inner surface of the two hemispheres, the limbic system is composed of the structures identified by Papez plus several other centers beneath the cortex, including the *amygdala*, the *septal areas*, and the *prefrontal cortex* toward the extreme front of the brain (see diagram 6). When electrodes are placed in some of these areas, animals will show all the outer signs of rage. Stimulation in other areas arouses pleasurable responses, indicated by the animals' enthusiasm to stimulate themselves over and over again by pressing a bar activating the stimulating electrode. Sometimes the animals activate what came to be called the "pleasure center" as often as 7,000 self-stimulations an hour.

Summarizing a great deal of experimental work, it's fair to say that, depending on the areas stimulated, the limbic system serves as a generator of agreeable-pleasurable or disagreeable-aversive affects. What's more, these affects and their subsequent emotional expression can be produced by electrical stimulation in humans. Here is a description of a woman suffering from epileptic seizures originating in the limbic system:

> Her main social problem was the frequent and unpredictable occurrence of rage, which on more than a dozen occasions resulted in an assault on another person with a knife or pair of scissors. The patient was committed to a ward for the criminally insane, and electrodes were implanted in her amygdala and hippocampus for exploration of possible neurological abnormalities. They demonstrated marked electrical abnormalities in both amygdala and hippocampus. It was demonstrated that crises of assaultive behavior similar to the patient's spontaneous bursts of anger could be elicited by radio stimulation in the right amygdala.

Considered from the point of view of size alone, the amygdala occupies an extremely small proportion of total brain volume (this is undoubtedly one of the reasons Papez ignored its importance when formulating his "circuit"). An early neuroanatomist thought the small organ looked like an almond and named it from the Greek and Latin words for the almond (L. *amygdala*). But despite its small size, it plays a key role in the channeling of drives and emotions. We know this partly from observations of patients like the one just described and partly from experiments on animals.

In one particularly fascinating study with monkeys, the researcher cut the optic chiasm, that part of the brain where some of the visual fibers from each eye cross over and pass to the visual area on the opposite side of the brain. As a result of this operation, visual impulses from each eye traveled only to the visual area on the same side of the brain. Next, the researcher destroyed the amygdala on only one side. Normally visual impulses from each eye on their way to the visual cortex connect to the amygdala on the same side. But as a consequence of destroying the amygdala on one side, only one eye could convey visual information to the remaining intact amygdala on the same side. The experimenter then put a patch on each of the two eyes in succession and approached the animal.

As a rule monkeys when confronted with humans react with hostility and aggression. And this is what occurred when the experimenter applied the patch to the eye on the side of the destroyed amygdala. But if he patched the eye on the side of the intact amygdala, the monkey failed to react and in place of aggression merely sat placidly eying the experimenter. In each instance the visual event was the same—the experimenter approaching the monkey—but the customary aggressive response only occurred when the experimenter was viewed through the eye on the same side as the intact amygdala. It was as if the amygdala provided the emotional coloring and interpretation of what the animal was seeing. From this early experiment carried out over thirty years ago researchers concluded that the amygdala is an essential component for linking emotional responses to happenings in the environment.

Another clue pointing toward the amygdala's importance in emotion came from observing the results of destroying the amygdalae on *both* sides of the monkey's brain. Named after the two brain researchers who first described it, the Kluver-Bucy syndrome involves three striking and bizarre behavioral responses. The monkeys no longer express aggression toward their human captors; they indiscriminately try having sex with anything they come into contact with; they lose the ability to distinguish by vision alone edible from inedible objects and keep mouthing everything they come into contact with. The common element in all of these seemingly unrelated behaviors is a breakdown in the channeling of drives and emotions to their appropriate external visual targets.

More naturalistic experiments carried out in the wild also support the idea that the amygdala is a center for a wide range of emotions and behaviors. For example, if electrodes are inserted into the amygdalae of free-ranging monkeys and the electrical activity monitored at a distance by radiotelemetry, increased discharge rates are recorded during fights and sex. If the electrodes are removed, the amygdalae destroyed, and the animals returned to the wild, they cease to act like the other monkeys. Their behavior no longer conforms to their position within the dominance hierarchy within the colony; they lose the ability to pick up the social signals that bind the members of the colony into a unit; and they dispense with traditional rituals like mutual grooming, effectively isolating the offending monkey from the other members of the colony. The vocalizations of the amygdala-damaged monkey also lack the appropriate emotional overtones and, as a result, the affected animals are ignored or, more often, attacked.

"These observations in monkeys suggest that the amygdala, together with other limbic components, may be essential for the interpretation of emotion-conveying gestures and vocalizations and also for the ability to emit these in the proper context," according to neurologist M. Marcel Mesulam.

While it is difficult identifying the anatomical and functional organization of emotions within the brain, it's even harder isolating the separate contributions made by the different limbic components. For one thing, emotions are hard to recognize and define. We recognize that some people are more "emotional" than others; it's usually difficult to be more precise. Like someone once said about adolescence, "I can't tell you exactly what it is, but I know it when I see it."

For years neuroscientists sought a method for identifying and localizing a single emotion within the animal brain. Their greatest difficulty was in determining what emotions animals actually experience. Since our emotional experience is heavily dependent on language, and animals can't talk, this places us at a particular disadvantage in identifying their emotions. Indeed, we often don't know for certain the emotion we are experiencing ourselves until we can come up with a name for it. For instance, if epinephrine (adrenaline) is infused intravenously into your arm, it increases blood pressure, speeds up the heart rate, and quickens respiration. Depending on the circumstances surrounding the infusion and what you have been told beforehand to expect, you may experience and describe anxiety or, instead, calmly register the ensuing physiological changes without reporting any emotional experience at all.

Fear is a particularly appealing emotion for neuroscientists to explore because of its universality. One doesn't have to devote a lifetime to the study of animal behavior to observe that animals respond to threat with specific emotional responses. The snarl of our dog or the arched back of our cat or the puffing of the neck feathers of our parrot occur when something is happening that the animal believes it has reason to fear. Among animals commonly employed in laboratory experiments, the rat will suddenly "freeze" and its blood pressure go up under conditions of threat. Moreover, these responses can be conditioned by classical conditioning experiments.

In the artful experiments of Joseph LeDoux of the Center for Neural Science at New York University, a rat is exposed to a simple sensory stimulus, like a tone followed by a brief footshock. After a few repetitions of this pairing of tone with footshock, the tone alone is sufficient to elicit the "freezing" and elevation of blood pressure. After this, LeDoux then begins delicate brain dissections aimed at revealing the relevant areas involved in the fear response.

The tone is conveyed from the ear along the auditory pathway to the auditory cortex in the brain's temporal lobes. Along the way, the fibers synapse (make contact) at a way station within the thalamus known as the *medial geniculate*. From here, two pathways diverge to the auditory cortex. The largest proportion of fibers from the medial geniculate project to the auditory cortex. Information is then conveyed to the association areas for additional processing and relay on to the temporal and pre-frontal association areas which connect directly back to the amygdala. This corticoamygdalar pathway involves several stops along the way.

A second fiber pathway from the medial geniculate, the thala-moamygdalar, does not ascend to the cortex but passes directly to the amygdala.

An unexpected finding of LeDoux's research is that complete destruction of the auditory cortex, the part of the brain concerned with the reception of sound and the receiving point for the direct impulses from the amygdala, had no effect on the conditioning of fear to the tone. I say "unexpected" because the most intuitive and seemingly commonsensical interpretation of the conditioned fear response would be that the sound must be delivered to the auditory cortex first (how can you be afraid of a sound if you haven't heard it, right?) and then, after additional processing, in the association areas and elsewhere in the cortex where fear is aroused.

But although the auditory cortex is not required for fear conditioning, the amygdala is. If the lateral aspect of the amygdala is destroyed, the conditioning response to fear in the rat no longer takes place. Destruction of the medial geniculate body, the way station for the auditory impulses on their way to the auditory cortex, has the same effect, suggesting to LeDoux that connections between these subcortical structures form the basis for the fear response. This means that fear, at least the conditioned fear in the rat, occurs below the cortex.

"Together these findings indicate that the amygdala is a key structure for fear learning," LeDoux told me during a discussion. He mentioned several important implications of this discovery. A tone or other sound reaching the amygdala directly from the medial geniculate undergoes far less processing than one that ascends to the cortex for later reprojection back to the amygdala. The speed of transmission is also faster. Thus the thalamoamygdalar pathway is ideally suited for processing emotions and emotional responses when it is important to respond quickly to a possible threat even before the complete stimulus has been processed by the brain.

A sound in the brush of the Serengeti Plain may indicate a lion or merely the play of the wind. But a response is called for prior to a full determination of these alternatives. It is the thalamoamygdalar pathway that does this. Moments later the fear response may be further increased when the presence of the lion is confirmed on the basis of more complete information reaching the amygdala from the cortex (the corticoamygdalar pathway). We encounter here another example of the brain's parallel processing capabilities. In one circuit, the emotional responsiveness must be quick even in the face of uncertainty. A false positive (treating a harmless stimulus as potentially threatening) carries less risk to survival than a false negative (treating a potentially threatening stimulus as harmless). In the second circuit, the basis for fear is fully elaborated and the resulting response to it can be shaped by conscious deliberation.

Responding on the basis of nerve cell connections that bypass the cortex implies that much of our brain's emotional processing occurs unconsciously and that the medial nucleus-to-amygdala circuit underlies our "unconscious" fear responses. That means that we as conscious creatures can and do react with fear even though consciously we haven't the slightest idea what it is that is frightening us. Heightened "startle responses," sudden onsets of anxiety or even panic, personality characteristics like being "hyper" or "edgy"—these are everyday examples.

As opposed to this "unconscious" processing, the activity in the corticoamygdalar circuit would seem to imply a fully conscious processing of emotions. But I am skeptical about the possibility of our ever being completely conscious of our emotions, and LeDoux provides good reasons for my skepticism.

While the existence of an emotional processing circuit that bypasses the cortex strongly argues for unconscious emotional processing, it is likely that even when the amygdala is activated by cortical sensory systems the processing occurs unconsciously. When we are conscious of processing, we are aware of the consequences of processing after the fact rather than being aware of the processing itself. . . . Emotion, like other information processing functions, is processed unconsciously and only the consequences of processing are represented as conscious intent, and only sometimes.

In short, our belief that we "consciously" determine our fear responses is only a reincarnation of the over-esteemed homunculus bequeathed to us for perpetual care by Descartes. But as LeDoux's research makes clear, the emotional "meaning" of a sound is not processed within the cortex by some "supervisory" structure or neuronal arrangement, but at the subcortical level, even before the sound has reached the auditory cortex. Indeed, this finding that fear is organized within subcortical memory circuits explains in part why phobias and other fear and anxiety emotional disorders are so difficult to treat by psychotherapy and other "talk" therapies. Although the patient may talk freely about his phobias, recognize their irrational nature, and take determined steps to control them, they can return in full force in response to some seemingly trivial event. That's because that event originally activated the fear response at the level of the thalamoamygdalar pathway. At that level, the phobia is maintained intact, despite talking about it, denying it, willing it away, or other acts of disavowal by the cerebral cortex. Indeed, the intransigence of phobic symptoms in the absence of drug treatments indicates that the brain's memory for emotional experiences is an enduring one embedded within the thalamoamygdalar pathway. This is the reason deep-seated hatreds and fears on the part of one racial or ethnic group toward another are so difficult to modify by persuasion. This persistence of ethnic hatred does not excuse us from exerting every effort to protect the human rights of the persecuted and exploited, but it does mean we cannot expect long-simmering and deeply seated hatreds and feuds to disappear overnight. Those subcortical circuits involving the amygdala particularly arouse emotional passions at levels inaccessible to conscious or willed deliberation.

Societal violence is the most visible and important consequence of uncontrolled fear, hatred, and rage. Recently brain research on the limbic system has been applied as an "explanation" for individual violence. At issue is the question of whether or not subcortical circuits for the emotional expression of violent impulses may so overwhelm the cerebral cortex that the individual cannot resist violent impulses towards others. Indeed, the discovery that emotions can be turned on and off by brain stimulation in the limbic system and that some violent people act violently on the basis of dysfunctions of the subcortical brain areas introduced a new way of looking at aggression and violence.

While most neuroscientists agree that emotions are mediated by the limbic system, there is little agreement in regard to the individual contributions to emotion of the different components, and the influence that each component exerts on the others.

In animals, multiple areas within the limbic system increase, modify, or inhibit aggression. Moreover, the same area may enhance or diminish aggression depending on the site selected, the conditions of the experiment (electrical stimulation versus destruction of brain tissue), and the animal selected for experiment. Destruction of one limbic system component, for instance, the cingulate gyrus, increases aggression in dogs and cats, while the same operation in monkeys and humans exerts a calming action.

In people, aggressive and violent acts tend to be associated with abnormalities such as tumors or birth injuries in the frontal lobes, the amygdala, and the temporal cortex and accompanying areas of the limbic system connected to it.

With the development of the electroencephalogram (EEG; brain wave test) in the 1940s and 1950s, neurologists achieved a major insight into the relationship between abnormal electrical discharges in the brain and behavioral disturbances. During an epileptic attack, the EEG showed electrical discharges within various areas of the brain, depending on the location within the brain of the abnormality (tumor, scar, developmental anomaly, and so on) responsible for the seizure. Most intriguing were the EEG findings in those forms of epilepsy marked by such things as confusion or staring or aimless wandering, now termed complex partial seizures or psychomotor seizures. In these cases, too, seizure discharges recorded from the EEG correlated with behavioral disturbances. But the correlation was far from perfect: Many patients who behaved in the same way as patients with complex partial seizures had normal EEGs. The solution to this conundrum was provided in the 1960s by neuropsychiatrist Robert Heath and others.

Using implanted electrodes, rather than relying on electrodes attached to the scalp, Heath demonstrated seizure discharges occurring within the limbic system of patients with perfectly normal EEGs.

"These studies reveal that forms of electrical activity can occur in the limbic system without reflection in nearby structures or on the temporal cortex, let alone in scalp recordings, and that behavioral changes associated with this limbic . . . activity are seldom characteristic of grand mal or psychomotor epilepsy. Such observations remind us that the clinical EEG is an extremely "blunt diagnostic instrument," according to psychiatrist Russell Monroe.

In 1974 Monroe proposed that a specific form of violent behavior was the result of a "limbic ictus" (seizure). Episodic dyscontrol as defined by Monroe consists of behavioral changes which are ". . . intermittent or paroxysmal, demonstrating an abrupt onset and an equally precipitous remission. They experience intense dysphoric [unpleasant] affects associated with impulsive action. They describe symptoms suggesting a prodrome or aura [warning] associated with impulsive actions that might be considered *formes frustes* of epilepsy."

Episodic dyscontrol in a fifty-one-year-old father of three who worked in a middle management executive position, for example, took the form of sudden facial flushing, heart pounding, "sparks" before his eyes, and the welling up of rage. He felt totally out of control and would begin smashing windows, breaking up household furniture, punching the walls and doors to the extent of fracturing his hand. On occasion his fury led him to make reckless, impulsive excursions in his car that ended in minor accidents. Finally, as the fury abated, he would return home contrite and embarrassed.

Monroe insisted that episodic dyscontrol did not refer to just any violent act. "The magnitude of the behavior during an episode is grossly out of proportion to any precipitating event. The individuals often describe these episodes to 'spells' or 'attacks.' The symptoms appear abruptly, and regardless of duration, remit almost as quickly. Following each episode there is usually regret or self-reproach at the consequences of the action and the inability to control the impulses. Between episodes there are no signs of generalized impulsivity or aggressiveness."

Nor is episodic dyscontrol associated with lifelong personality disorders. Mostly the episodes are atypical of the person's usual behavior and character. During the episode, the person may seem confused or somehow out of contact. Following the episode, partial or even complete amnesia often occurs. The behavior is often inexplicable to others. "Even the individual affected is often startled by his own behavior, describing the events as resulting from a compelling force beyond his control, even though he is willing to accept responsibility for his actions," according to Monroe.

Other commonly encountered features include a hypersensitivity to loud sounds or bright lights. And although epileptic seizures are rarely present, minor EEG abnormalities and "soft" neurologic signs and symptoms reflecting limbic system involvement are frequently found.

Patients displaying violent outbursts of aggression without any apparent trigger are also more likely to come from low socioeconomic levels and to have experienced difficult births, often associated with brain damage. As children, many are hyperactive and accident prone. As adults, they continue this tendency in automobiles and accumulate frequent citations for traffic violations, serious accidents, and aggressive, even assaultive, use of their automobiles. Many display symptoms of pathological intoxication—responding with violent outbursts after even small amounts of alcohol.

In the 1970s and 1980s, University of Pennsylvania neurologist Frank Elliott studied the frequency of neurologic abnormalities in 286 patients with a history of recurrent attacks of uncontrollable rage occurring with little or no provocation. Objective evidence of developmental or acquired brain defects turned up in 94 percent of the cases.

With the establishment of episodic dyscontrol as a form of brain disease, other forms of disruptive and violent behavior came under scrutiny. This led to a spate of defense claims throughout the 1960s and 1970s that various *nonexplosive* forms of violence might also be based on a brain abnormality. (Jack Ruby's killing of Lee Harvey Oswald was probably the most famous attempt at using this defense. It failed.) But eventually fewer of these claims reached the courts because, with the exception of episodic dyscontrol, a relationship between violence and brain "disorder" or "damage" cannot be reliably established in a specific instance.

But far more is involved here than a mere quibble between neurologic and legal experts regarding legal standards of proof. If widespread social violence can be attributed to brain disorders, then many of our attitudes toward individual responsibility come up for question. No less is at stake than a radical reevaluation of the traditional concept of free will.

"The fact that feeling states may be induced by defects in the limbic area even at complex levels of perception opens up for question whether genetic and environmental factors acting on brain physiology are not finally determinative of many acts of deviant behavior," writes lawyer-psychiatrist Lawrence R. Tancredi in *International Journal of Law and Psychiatry*. He goes on to suggest that "With a re-examination of the delicate balance between mind and body—that is brain as tissue influenced by environment, genetics, and biophysiology, and mind as that cognitively operating part of the personality which involves individual control—we may discover that a vast majority of individuals engaging in deviant behavior are influenced by factors and determinations which are clearly not within the powers of their individual free will."

In a phrase, the neurologic defense resolves the traditional tension between the bad and the mad by redefinition: the mad and the brain-damaged, or (as one expert refers to them) the "biophysiological defectives."

But before assenting to this brave new world in which the determination of individual responsibility is given over to neurologists and neuropsychologists and the concept of free will is relegated to the status of a museum piece dating from medieval scholasticism, consider the following:

Brain damage and brain deficit are not the same. "Damage" refers to deviations from normal structure; "deficit" refers to behavioral abnormalities, specifically: actual impairment, loss, or alteration of intelligence, emotion, or the capacity to act freely.

Nor is there any necessary correlation between damage and deficit. A host of factors in modern life can lead to brain damage, but not necessarily to deficit: birth difficulties; childhood illnesses; the after-effects of drugs; exposure to toxins; accidents and physical abuse resulting in head injury. Any or all of these, alone or in combination, may produce test results suggesting some degree of "organicity" or "brain damage" yet exert little or no discernible effect on intellectual or emotional functioning. Even in those instances when damage leads to deficits, the range of expression following similar brain injuries is enormous. That's because no two human brains, even those of identical twins, are exactly alike. Nor can the brain's response to the same trauma be predicted.

The human brain is marvelously resourceful and often able to compensate for injury or "damage" without exhibiting deficits. This is particularly true when it comes to the control of aggressive impulses.

Moreover, total unanimity does not exist about what constitutes a "normal" brain. Deviations from "normal" routinely occur in tests of brain function and structure. At the minimum, somewhere between 3 and 5 percent of people drawn at random may demonstrate abnormalities in their electroencephalogram. Magnetic resonance imaging (MRI; a method of demonstrating brain structure) routinely turns up anomalies. Yet these deviations seem to exert no discernible effect on normal functioning.

Do such abnormalities represent brain damage? Perhaps in the narrow neurologic sense; but certainly not in the legal sense. In general, the law requires that three criteria be met for alleged "brain damage" to serve as a mitigating factor in determining responsibility.

First, does "brain damage" exist, and is it responsible for a deficit? Second, is that deficit a contributing cause of the defendant's action? Third, can it be said convincingly that absent the deficit and the resulting behavior, the crime would not have taken place?

As to the first criterion, electroencephalographic recordings suggest that the prevalence of EEG abnormalities among violent people ranges from 25 to 50 percent, compared to 5 to 20 percent among normals with no history of violence. The percentage of EEG abnormalities is even higher among subjects who have committed violent acts on more than one occasion. And a study of fifteen death-row inmates—conducted by a psychiatrist and a neurologist who spend much time in court advocating the neurologic defense—found that ten inmates had evidence of neuropsychological abnormalities and all fifteen gave a history of head injury. The same psychiatrist and neurologist in another study found that of thirty-seven death-row prisoners who had committed capital crimes while juveniles, eleven were reported to have suffered head trauma; nine had abnormal EEGs or neurologic examinations; and twelve gave a history of physical or sexual abuse.

"The reasonable conclusion from these neuropsychiatric studies of death-row inmates is that both juvenile and adult groups have significant findings," wrote psychiatrists Gregory B. Leong and Spencer Eth in *The Bulletin of the American Academy of Psychiatry and the Law.*

So far, so good. But when one comes to the second, and especially the third, criterion, things begin to break down.

For one thing, the great majority of people who commit violent crimes have never been demonstrated to be suffering from "brain damage" or any of its alleged consequences. Furthermore, in most instances there is nothing in their background or the details of their crimes which even raises the question of "brain damage." In many instances, cultural factors are preeminent. Indeed, within certain cultures violence and murder are perfectly acceptable means of settling differences and discharging aggression.

Within our inner cities, violent attacks and murders occur every day for reasons that appear tragically inconsequential to most of us. Are we to assume that when one person kills another because of a perceived lack of respect ("dissing"), that the perpetrator suffers from some form of "brain damage"? Are we to assume that the epidemic of drive-by killings, in which people chosen at random are shot to death, can be traced to "brain damage" instead of a culturally linked expression of violence?

As an additional complicating variable, "brain damage," despite its seeming precision, is a term only slightly less subjective than the "insanity" it is intended to replace. For instance, a slightly abnormal electroencephalogram may represent "brain damage" to one neurologic expert who, for reasons having less to do with neurology than with personal ideological convictions about the death penalty, may choose to ignore the fact that somewhere between 2 and 4 percent of electroencephalograms done at random on members of the general population turn out abnormal. Still other experts rely strictly on neuropsychological tests, despite the fact that spirited controversies presently exist among neurologists and psychologists as to how such tests should be interpreted and how much emphasis should be placed on abnormalities when they do turn up.

Finally, while episodic dyscontrol and its variants are generally accepted by most experts on violence as legitimate contributors to diminished responsibility on the part of persons who kill, there is no agreement within the psychiatric or neurologic community that childhood physical or mental abuse, head injury alone, or abnormalities found on neuropsychological tests—three factors especially favored by expert witnesses for the "neurologic defense"—necessarily imply any appreciable impairment in a person's ability to control and contain violent impulses toward others. This is especially true when it comes to crimes that involve premeditation or are carried out with forethought and were planned over extended periods of time.

There is yet another reason to doubt any facile assertion that a violent act is attributable to brain damage. In general, the more complex and symbolically meaningful a behavior or emotion, the less likely it is that it can be linked directly to specific brain areas. This is particularly true when it comes to what might be called the "major" emotions— love, hate, loneliness, aggression, among others—which are mediated by many brain areas and the interaction of billions of nerve cells and thousands of neurotransmitter interactions.

Despite these limitations in our knowledge about the brain and its relationship to violence, courts are increasingly willing to accept "brain disease" as a mitigating factor in determining guilt or innocence. This has stimulated a good deal of speculation and research into the existence of a "criminal brain": alterations in brain anatomy or functioning responsible for criminal behavior.

Slowly, the emphasis is shifting away from the principle that a person is responsible for his or her behavior and toward various "explanations" why certain people engage in criminal or other self-destructive actions. On the face of it, this redefinition of free will and individual responsibility seems to make sense. After all, if a person can suddenly become violent as the result of an epileptic seizure or other brain abnormality, why can he not commit crimes that appear to involve premeditation, such as with serial killers or stalkers, for the same reason? With this question, brain scientists come to grips with the issues of good and evil.

What becomes of our traditional belief in personal responsibility if the killing of another person is viewed not as a matter of choice but, rather, as due to some irresistible impulse emanating from a damaged brain? And if the extremes of human destructiveness emanate from a disturbance of the brain, does that necessarily imply that the exemplars of human goodness—the Mother Teresas of the world—also act the way they do because of a certain organization within the brain? If acts of violence and savagery are the result of a given pattern of brain organization, then it seems to follow that goodness and acts of kindness must also be judged as somehow determined for us by neurological factors. Although I am not comfortable with this neurological predeterminism, I must admit in many instances it seems true. With some people, acts of consideration and kindness towards others seem natural, indeed even inevitable. It is as if they couldn't imagine themselves acting any other way. If this is true, what happens to free will? Are such people acting kindly only as a result of some patterning within their brain? Are those who love and those who hate others merely acting out different brain activity patterns? I recognize that in raising such questions, I am proceeding quite a bit beyond my own training and education. I am neither a theologian nor a moral philosopher. But that aside, it does seem to me that a belief in goodness and the existence of good people who have loving and caring feelings toward others must imply a belief in the existence of evil or whatever word you might wish to substitute for people who not only commit, but seem to enjoy committing, gratuitous and inexplicable acts of cruelty and destructiveness toward others.

I'm thinking now of a twelve-year-old boy I examined some years ago who had murdered another child. He told me of sneaking up on the child who was swimming in a stream. As he recounted how he had bashed in the other boy's head and face with rocks thrown from the shore, a cold and eerie smile flashed just briefly across his face. When I asked him about this unexpected and chilling display of amusement, he elaborated on how the other child had cried out for his life and begged his attacker to spare his life. Again the smile played across his features. This act of heinous cruelty and barbarity had obviously given him great pleasure and continued to do so. Brain damage? All of the tests of brain function turned out normal. Insanity? He suffered from none of the hallmarks of presently recognized mental illness and he certainly did not experience anxiety, remorse, or depression for what he had done.

To other people, violence comes so naturally that they can't express themselves any other way. An example of such a person was a Correction Officer who headed up a "goon squad" inside a prison for the criminally insane. "We were there to counteract violence by escalating it. If any of the prisoners acted up we showed them that we could be even more violent," he told me in answer to my question about how he saw the purpose of his work. This man, a former marine and private security guard, came under my care after he barely escaped being killed by prisoners who threw him from the second tier of the cell block into the waiting clutches of another prisoner who attempted to hang him with a bedsheet. Only at the last second was he rescued by other guards. Now retired as a result of the injuries received, this man's life continues to revolve around violence. "If I see a violent movie, I want to get dressed and walk the streets in the toughest section of town, just hoping somebody tries to attack me."

When listening to a man like this, it's difficult to escape the conclusion that violence is written into the very circuits of his brain. Even when he wants to act peacefully, he must make special efforts to control his preoccupation with violence. "You seem like a nice enough guy, Doc," he said to me one day in the middle of a perfectly unstressful conversation. "But even now I'm thinking that maybe I want to take a swing at you."

How does one conceptualize such propensities toward violence and other examples of loss of emotional control? Joseph LeDoux: "It is easy to imagine how genetic variation in the extent to which thalamo-amygdala and cortico-amygdala circuits function independently could predispose individuals to different degrees of control of emotional behavior and to differences in their sense of being in control."

But legal determinations are made not on the basis of probabilities or possibilities but on what can be stated with confidence in regard to one specific instance of violence. And no matter how much we learn about the brain and those brain areas involved in violence, we can rarely answer that question in the individual case. That's because we cling to several incorrect assumptions about the brain and behavior.

First, the traditional distinction between mind and brain embodied in the two often-competing disciplines of psychiatry and neurology is now untenable. All of the expressions of mind (the exercise of free will, memory, learning, emotional expression, and so on) are now recognized as affected by the state of the brain. The brain in turn is influenced by genetics, diet, and the social and economic environment, which in those subcultures marked by increased violence frequently involves poor prenatal care, infections, head trauma, and other forms of physical and emotional abuse.

Second, neurological explanations of human behavior are rarely determinative. Social, economic, and genetic factors are usually more contributory to violence than brain abnormalities or defects. Despite these other contributors, alleged neurological explanations for violence run the gamut from clear-cut instances when the violent act appears to have been neurologically inevitable to instances at the other end of the continuum when a brain basis for the violence is only one among many possible explanations.

However we might wish it otherwise, neurology is not going to solve the mystery of why some people kill others. Neither can it help us discover why killers are often not just unwilling participants in something beyond their control, but rather, judging from their own words and actions, often engaged in something that gives them great pleasure. As Ronald Markham, who has examined more murderers than perhaps any psychiatrist in the United States, says: "Our society is leaning awfully close to the idea that you have to be mentally ill in some way to commit a crime. This is not so. Most crimes—even grisly murders—are not committed by mentally ill people, but by people just like you and me."

It's likely that the tendency towards violence, like most human behaviors, follows a bell curve. At one end are those who, even in the face of extreme or life-threatening provocation, cannot arouse themselves to violent action (the conscientious objector). Further along the continuum are the rest of us, who are capable of violence if the stakes are high enough. At the other extreme are the habitually and chronically violent, whose actions do not represent insanity and certainly not brain damage, but only the outer limits of our human potential for violence.

Violence and other emotional expressions of the human brain contradict our notion of ourselves as "rational" creatures whose actions are determined solely by logic and reason. And as we have seen, emotions are not always pleasant to encounter in ourselves or others. But it's not my intention to overemphasize the negative aspects. Emotions can also stir us to great efforts and great creations. In fact, emotions form the underpinning for our most transcendent experiences and dazzling creative efforts.

Mystics, poets, saints, seers and lovers describe their inner experiences in highly emotional words and phrases. Nietzsche captured this emotional side of creativity: "You can have no idea of the vehemence of such composition," he wrote to his sister about his masterpiece *Thus Spoke Zarathustra*. And in *Ecce Homo* he described the emotional nature of creativity: "Has anyone . . . any distinct notion of what poets of a stronger age understood by the word inspiration? . . . There is an ecstasy such that the immense strain of it is sometimes relaxed by a flood of tears, along with which one's steps either rush or involuntarily lag, alternately. There is the feeling that one is completely out of hand, with the very distinct consciousness of an endless number of fine thrills and quiverings to the very toes. . . . Everything happens quite involuntarily, as if in a tempestuous outburst of freedom, of absoluteness, of power and divinity."

Nietzsche puts into words what every creative person experiences. But he said little that makes sense today about the origins of the creative impulse. Is the brain of the creative person organized differently from his or her less gifted counterpart, or is the difference one of degree—perhaps something like a more efficient use of resources? When it comes to the brains of gifted, creative people—superior brains, if you will permit that somewhat elitist-sounding term—does the modular concept fit here as well?

CHAPTER TWELVE

The Illness of Maurice Ravel

S urprisingly, very little has been written on the brains of highly creative people. For one thing, neurologists and neuropsychologists—
the two specialists who have contributed the most to our
understanding—are usually occupied with the diagnosis and treatment
of disorders. Rarely do they have the opportunity to explore the neurological basis for exceptional mental ability or creativity. Most instances
have involved highly gifted people who are rendered less creative as a
result of brain damage. The composer Maurice Ravel provides a tragic
confirmation that music is organized within the brain along the lines of
the modular concept we have discussed throughout this book.

At fifty-eight years of age, Ravel was afflicted with a progressive neurological illness. During the final four years of his life, he completely lost
the ability to compose, despite retained abilities to listen to and remember his own compositions (until late in the illness he could point out
subtle departures in performance from the scores of his compositions as
he had originally written them). The sequence leading up to Ravel's
disability is intriguing and suggestive.

In 1933 a pupil of Ravel's commented that the master was making
frequent spelling mistakes on the scores of his pieces. Similar spelling
errors turned up in his letters. When this tendency was gently pointed
out, Ravel expressed surprise, but could do nothing to reverse the
process, which progressed rapidly. By the year's end he could no longer
sign his name or read. In regard to speech, he had difficulty with proper
names but understood others quite well and expressed himself clearly
enough to be understood. During this period he continued to compose
without any obvious diminishing of his abilities.

As with his earlier spelling mistakes, which heralded more serious communication difficulties to come, Ravel remained largely unaware of earliest signs of the insidious process that would rob him of his musical gifts. In September 1933, Ravel spoke to an interviewer about plans for his opera *Jeanne d'Arc*. Two months later he had made no progress on the work and confided to a friend: "I will never write my Jeanne d'Arc; this opera is over. It's over, I can no longer write my music." Ravel's descriptions of his difficulties are illuminating.

Although he could still generate music "in his head," he could not write, play, or sing it. He could not sight-read a score or play by heart more than a measure of his own music. Despite these difficulties, he had no problems playing major and minor scales on the piano and retained an accurate memory for his own music when hearing it played by others.

Ravel's illness was most likely the result of a localized cerebral degeneration involving two distinct but geographically contiguous structures sustaining, respectively, verbal and musical reading functions. But whatever the exact details concerning the cause and location of the brain damage, Ravel's illness, according to Justine Sergent, a psychologist at the Cognitive Neuroscience Laboratory of the Montreal Neurological Institute, consisted of "a selective impairment of functions underlying the translation of musical representations from one modality to another. Ravel had become musically illiterate, not because he had lost musical knowledge or technique as such, but because he was no longer able to use this knowledge in an integrated manner in order to translate musical expression from one modality to another."

Ravel's disability is doubly sad because all of the elements of his genius remained intact; he simply was no longer capable of the integration and coordination required for musical composition. This isolation of the respective modules reminds me of the people we encountered in Chapter Eight who suffered from frontal lobe injuries. They too retained all of the necessary components for appropriate behavior, yet were unable to put everything together, to link up the modules in an integrated fashion.

Ravel's experience also suggests that disturbances in the brain processes underlying subtle and highly evolved functions may escape detection; damage to the brain of a musician of lesser genius than Ravel might even go unnoticed. If you believe, as I do, that the brain of each of us is unique—no other brain now or at any time in the past is exactly like my brain or your brain—then tiny injuries may bring about changes that are recognized only by the affected person himself.

Tragic instances such as Ravel remind us that in order to understand and nourish creativity, it is not enough to concentrate on social, psychological, and cultural contributions. Although these factors are important, we mustn't lose sight of the ultimate determinant of creativity: the human brain.

Recent research reveals that various aspects of creativity are mediated and nourished by specific organizational patterns within the brain. This is not the same thing as saying that creativity can be explained on a strictly neurologic basis—a form of reductionism Arthur Koestler rightly criticized as "nothing buttery"—but it does suggest that some understanding of the brain may contribute toward that desire for personal creativity.

When the two brain hemispheres are compared, the right seems particularly important in the visual and performing arts. Numerous studies on normals, neurosurgical patients, and brain-injured people reveal that the right hemisphere is generally dominant for recognizing and identifying natural and nonverbal sounds. It is better at appreciating depth perception; maintaining a sense of body image; producing dreams during REM sleep; and appreciating and expressing the emotion aroused by music and the visual arts. Finally, the right hemisphere is specialized for perceiving emotional expression in others and generating it in oneself.

Damage to the right hemisphere results in distortions in the appreciation of music and natural sounds. Justine Sergent, whose study of Ravel I mentioned previously, carried out a PET scan study of musicians engaged in sight-reading (deciphering an unknown musical score and playing it on an instrument). In contrast to the reading of words, the sight-reading of a musical score relies on spatial information, i.e., the position of notes in relative position to each other rather than simply their individual position on the staves. The features of individual notes are thus less important in reading music when compared to the distinguishing features of the individual letters in reading words. Another difference between sight-reading music and reading words is the translation that takes place from the visual structural description of a score to an internal representation of the positions, patterning, and timing of the finger movements required in playing the music.

To learn what happens in the brain during sight-reading, Sergent employed a PET scan to illustrate the areas active while playing a scale from memory and during sight-reading. The final images were made by averaging the findings of ten professional musicians and then sub- tracting from the image the scale-playing condition The resulting image depicted the brain areas specific to sight-reading.

Sergent found three main areas of activation that occurred only with sight-reading and not with either scale playing or score reading alone. This indicates, says Sergent, that brain activation during sight- reading reflects operations by which musical representations are transformed from the visual and auditory modalities involved in read- ing and "hearing" the notes in one's head, to the patterning and posi- tioning of the fingers on the keyboard and, closely associated with that, the organization of the sequencing and timing of finger move- ments involved in playing a particular score that is being sight-read. In short, Sergent has found that different aspects of the musical expe- rience are mediated by different brain areas. If this is true, a break- down in one aspect of the musical experience could conceivably occur while leaving other aspects unchanged. This is exactly what occurred with Ravel.

Damage to the right hemisphere may cause, in addition to specific disturbances in the creation and expression of music and other arts, a host of neuropsychiatric expressions—all of them germane to creativity. These include indifference and loss of "drive," depression, manic excite- ment, euphoria, impulsivity, delusions, and perceptual distortions in regard to the body.

"The right hemisphere maintains a highly developed social-emotion- al mental system and can independently perceive, recall and act on cer- tain memories and experiences without the aid or active reflective participation of the left hemisphere," according to R. Joseph of the Neurobehavioral Center in Santa Clara, California.

Joseph is describing on the neurological level in somewhat technical language what every creative person experiences regularly: creative ideas and images seeming to spring "out of the blue," which might more accurately be characterized as out of the right hemisphere. "Of the truly creative no one is ever master; it must be left to go its own way," as Goethe described the serendipitous ways of creativity.

In all instances, personal insight into the source of one's own creativity is hampered by the fact that:

> We know more than we can say: we live
> in waves and feelings of awareness
> where images unfold and grow
> along the leafwork of our nerves and
> veins. . . .

as poet Peter Meinke describes the process in his poem "Azaleas."

With writers and other creative individuals who work with words, the right hemisphere contribution is undoubtedly much different. Since language and verbal centers are in the left hemisphere, writers are capable, at least in theory, of continuing to be creative in the face of right brain damage. What usually stops them is one or more of the emotional sequelae of right brain damage mentioned above.

Split brain patients provide an unusual opportunity to explore the mutual interactions between brain organization and creativity. These patients behave perfectly normally in ordinary social situations. But despite their apparent normality, they show little creativity as measured by tests of language, thinking, and emotional expression. They lack the ability to transform the imagery and symbols generated by the right hemisphere into creative verbalizations.

Dr. Klaus D. Hoppe, a psychiatrist at the Hacker Clinic in Los Angeles, believes that the absence of creativity in split brain patients is similar to what is observed in people suffering from *alexithymia*, a term taken from the Greek meaning "without words for feelings." People with alexithymia have great difficulty in identifying and verbalizing their feelings. And although the term is rarely encountered, alexithymics are far from rare. They appear frequently in medical clinics and doctors' offices. Typically they experience and express emotional stress by developing physical symptoms. Insight into the psychological origin of their distress eludes them. Moreover, they vigorously deny inner feelings and seem to inhabit a robotic inner life devoid of the finer shades of emotional experience. As a result, they are oblivious to suggestions that "stress" or other emotional factors may be playing a role in their illness.

Needless to say, alexithymics are difficult and frustrating patients to care for.

Creativity is the opposite of alexithymia. As a rule, creative individuals are "in touch" with their feelings and express them through their creative productions. At the neurological level, this involves an enriched communication between the hemispheres—*bisociation*—rather than the hemisphere dissociation typical of the split brain patient. The more creative one is, the more likely that the two sides of the brain are in easy communication with each other. In support of this are EEG studies carried out by Hoppe and associates: They found greater coherence and interhemispheric communication between the two hemispheres in creative people.

"It is now certain that the corpus callosum can transfer high-level information from one hemisphere to another," according to neurosurgeon Joseph E. Bogen of the University of California at Los Angeles, who carried out early research on split-brain operations in humans. Bogen suggests that the neurological underpinnings of creativity may require not only good communication between the hemispheres but "a partial (and transiently reversible) hemispheric independence," whereby one hemisphere may for a time independently engage in creative production outside of immediate conscious awareness. This could explain the "Ah ha!" response: "the illumination that precedes subsequent deliberate verification."

Anticipating such developments years ago, neuroscientist Frederick Bremer wrote that the corpus callosum, by uniting the activities of the two hemispheres, makes possible "the highest and most elaborate activities of the brain"—and what can fit this description better than creativity?

A good model of the creative process is suggested by Arthur Rothenberg, Clinical Professor of Psychiatry at Harvard medical school. He speaks of homospatial thinking: ". . . actively conceiving two or more discrete entities occupying the same space." To test this hypothesis, Rothenberg showed forty-three artists three sets of slide photographs; each set consisted of two different slides, one of five racehorses rounding the turn paired with a separate slide of five nuns walking together in Vatican Square. Some of the artists were shown the slides aligned side by side on the projection screen while others viewed the slides superimposed on each other. Each artist was then asked to create a pastel drawing stimulated by each of the three slide pairs. Two internationally prominent artists evaluated the drawings. They judged drawings stimulated by the superimposed slides as significantly more creative than drawings resulting from the separated slides.

Nor is homospatial thinking limited to drawing and the visual arts. Rothenberg believes creative people in literature, music, science, and mathematics excel in their ability to intermingle and superimpose elements from many different spatial and temporal dimensions. He calls this *Janusian* thinking: actively conceiving two or more opposites or antitheses simultaneously during the course of the creative process. These internal conceptions can be opposite or antithetical words, ideas, or images. After clarification and definition, they are either conceptualized side by side or as coexisting simultaneously. Finally, they are modified, transformed, or otherwise employed in creative productions in the arts, sciences, or other fields.

In a test of Janusian thinking Rothenberg administered timed word-association tests to twelve creative Nobel laureates, eighteen hospitalized patients, and 113 college students divided into categories of high and low creativity. The Nobel laureates gave the highest proportion of opposite (unusual) responses. They also furnished these opposite responses at significantly faster rates than common responses. Indeed, their average speed of opposite response was fast enough to indicate that conceptualizing the opposites could have been simultaneous. Rothenberg thinks of these highly creative people as engaging in a "translogical process."

Homospatial and Janusian thinking correspond in neurological terms to the actions of the frontal and prefrontal lobes. Thanks to these brain areas, which are more evolved in the human brain than in the brain of any other creature, we can mentally access information and keep it on-line (i.e., in mind) until it is integrated into one's ongoing plans. Thanks to this ability to bind time, we are able to hold on-line real or imaginal ideas that form the basis for creativity. We can internally rehearse and anticipate the consequences of our actions and introduce innovative and novel responses. We know this because, as mentioned in more detail in Chapter Eight, frontal and prefrontal damage reduces the injured person's world to the here and now, a world in which future consequences and possibilities exert little influence on present behavior and preoccupations. Such persons have great difficulty managing new situations or demands and almost never initiate innovative activities on their own. To this extent, frontal lobe disease and creativity are at opposite ends of a spectrum.

A high degree of cortical arousal (heightened responsiveness to the events and people in one's environment) is also a prerequisite for creativity. Since introverts characteristically show high cortical arousal, it's not surprising that, in general, creativity comes easier to introverts and those with emotionally responsive dispositions. "Creativity is a problem-solving response by intelligent, very active, highly emotional and extremely introverted persons," according to Dr. L. M. Bachtold, who over the past decade has concentrated on understanding the neurophysiological basis of creativity. His findings suggest to me that if there is a neurological basis for creativity, a single distinguishing trait separating the creative from others, it will be discovered within the frontal and prefrontal areas. "Subjected to a vast array of disorganized perceptual data and strongly feeling the inconsistencies, the active and intelligent individual forms new perceptual relationships to develop feelings of consistency and harmony," says Bachtold.

But there is a hazard in deriving one's neurological concepts about creativity strictly on the basis of observing the effects of deficits resulting from brain damage. It's always possible that the injured brain area may be only a small contributor to creativity, important enough to stymie the creative process but not at all its major impetus. Furthermore, there are rare instances on record where brain damage has *enhanced* creativity.

An artisan in his mid-twenties began to experience attacks marked by the feeling that he was floating helplessly in space. During these episodes, in which "waves" engulfed him, he would begin drawing impulsively. This behavior was all the more remarkable since he had never previously expressed any interest in drawing or artistic activity. An electroencephalogram showed epileptic seizure discharges within the frontal and temporal lobes on the left. Additional tests measuring brain activity levels yielded more evidence of dysfunction in the anterior regions of the left hemisphere.

Although a certain amount of speculation is always involved in explaining strange phenomena like this and, further, equally qualified observers may hold in good faith differing opinions about what's going on, I tend to agree with the opinion of the neurologists who cared for the patient. They speculated that the impulses toward artistic expression resulted from the release, as a result of damage to the left hemisphere, of the complex visual and spatial skills of the right hemisphere. It's as if blockage of this man's customary expressive powers brought on by left hemisphere injury led to an unusual and unpredictable form of artistic expression.

But despite the fascination engendered by this patient, it's necessary to remind oneself that this artisan is unique, the only example so far of enhanced artistic creation resulting from brain damage. Every other example I have personally encountered or read about in case reports confirms that brain damage produces a failing, and usually a serious failing, in creative expression. For example, brain injury to either hemisphere interferes with drawing ability in specific ways.

If the right hemisphere is damaged, the drawings lack features on the left side of the picture. This is because right brain damage results in neglect of persons and objects in the left visual field which is mediated by the right hemisphere. Spatial relationships are also distorted. Left brain damage, in contrast, results in overly simplified, childlike productions marked by conceptual rather than spatial errors (a drawing of a table setting for one may show three forks).

Probably the most feared form of brain damage results from Alzheimer's disease. And although the cause of this devastating degenerative brain disorder has so far eluded the efforts of neuroscientists around the world, a good bit is known about its effects on creativity.

An elderly artist underwent over a two-year period a dramatic deterioration in painting skills and intellectual ability. His artistic skills deteriorated slower than his other mental functions. In fact, until late in his illness, his artistic talents suffered only because of diminished motivation, memory, and organizational skills. In normal aging such deteriorations fail to occur.

A study of seventy-year-old patients carried out at the University of Gothenburg in 1989 showed that creativity did not decline over a thirteen-year follow-up period. The study was obviously an unusual one, since it involved tracking individuals already at an advanced age over more than a decade. The measures of reactivity were also somewhat unusual: the interpretation of ink blots coded for three components of creativity, fluency, flexibility, and originality.

Rather than declining with age, it's likely that creativity in the older person changes in nature according to the special circumstances of aging. For instance, self-expression for its own sake or as a sublimation for sexual or aggressive urges plays a smaller role. Like it or not, most of us really do mellow with age; even the most ambitious eventually share to some degree that ancient wisdom that recognizes, even in the face of fame and creative accomplishment, the transitoriness of all things. In addition, stylistic and thematic concerns are likely to change in the older creative artist in response to different interests than his younger counterpart. Picasso is an excellent example of the creative artist who retains an ever-evolving creative sense despite the ravages of time and aging.

Creative persons of different ages may also differ in the ways they go about artificially boosting their creativity. For centuries, some creative people have relied on chemical aids to spur their creativity. These chemicals include perfectly legal substances like tea (Balzac drank quarts of it during marathon writing sessions), alcohol (Faulkner and Kerouac drank heavily while in the process of composition), and nicotine (innumerable writers have attested to nicotine's power to enhance concentration and focus); also included are other chemicals that are either illegal or available only by prescription (amphetamines, cocaine, or LSD). The real question is, of course, do these chemical aids really stimulate creativity or are the users only fooling themselves?

In 1989 psychiatrists from the University of California at Irvine reported on a unique experiment carried out thirty years earlier on the effects of LSD on creativity. Artists were asked to draw and paint a Kachina doll. They then took LSD and drew a second doll. When they were finished, a professor of art history analyzed and compared the two productions.

The most significant changes occurred in the works of those artists with representational or abstract styles. Under LSD, the paintings were more expressionistic or subjective. In almost all cases the artists saw this as "fashioning new meanings to an emergent world." Indicators of this change included: relative size expansion, movement, alteration of boundaries and figure-ground relations, greater intensity of color and light, oversimplification, fragmentation, disorganization, and the symbolic and abstract rendering of people and objects. In short, there seems little reason to doubt that the creative product was different, but was it better?

Of course, this perfectly reasonable question is difficult to answer because of the highly subjective nature of art appreciation. Evaluating whether or not an artist is more creative as a result of taking LSD or another drug is not as easily done as deciding whether a scientist is more creative under the influence of psychoactive chemicals. With the scientist, one can perform experiments or check such variables as the number of papers written and how often they are cited by other scientists. But what type of measure does one employ for evaluating whether an LSD-influenced work of art is better than the artist's usual production? One thing is certain: No artist that I am aware of has created under the powers of LSD an acclaimed body of work over an extended period of time. Moreover, I can't help wondering if the artist's claims for the creativity-stimulating effects of LSD are not a variation of what we refer to in medicine as the placebo effect, where such psychological factors as expectations and hopes play an inordinately important role.

When it comes to psychological factors, the creative person must possess what psychologists refer to as *ego autonomy*. And various neurologic and psychiatric illnesses can impair this sense of autonomy. Obsession and compulsions interfere with spontaneity and the flexibility that is a requirement for creative breakthroughs. Frontal lobe disease, as mentioned a moment ago, disrupts programming, novel associations, and persistence. And whatever disease or malfunction interferes with smooth communication from one hemisphere to another prevents the integration and expression that forms the basis of all creativity.

But not all emotional and neurological illnesses are incompatible with creativity. Writers have a high prevalence of depression and bipolar disorder (manic depression). Since alcohol is the most commonly employed self-prescribed medication for depression, it comes as no surprise that alcoholism is also common among writers given to depression. It will be interesting to see if this pattern of alcohol abuse among creative writers will change in response to the current emphasis on health and the mild but definite social stigma attached to alcohol use. Also playing a factor in determining this will be the increasing acceptance of mood disorders as illnesses rather than character flaws, and the associated willingness of writers, artists, and other creative persons to seek help. As things now stand, some creative people refuse treatments such as lithium for bipolar disorder; they claim the medication blunts their sensibilities and lessens their creative powers.

Associations between neuropsychiatric illness and creativity have prompted some psychiatrists to conclude that a touch of madness may enhance creativity. While this is possible in the individual instance, most studies conclude that in general, creative individuals are most productive when their moods, thoughts, and behavior are under control. Certainly, untreated mental illness or brain disease can be expected to interfere with one quality of creativity identified by former world chess champion Max Euwe: "the ability to distinguish with certainty that infinitesimal dividing line between the inspired and the unsound."

Rather than correlating with brain and emotional disorders, creativity seems to be linked with what Freud referred to as ordinary human unhappiness. "We enjoy lovely music, beautiful paintings, a thousand intellectual delicacies, but we have no idea of their cost, to those who invented them, in sleepless nights, tears, spasmodic laughter, rashes, asthmas, epilepsies, and the fear of death, which is worse than all the rest," is how Marcel Proust described the psychological demands made on the creative individual.

One final point about writers. When thirty writers were administered psychological tests, their scores showed above-average IQs but failed to do any better than matched controls in any subtest other than vocabulary. This finding is further confirmation of the everyday observation that intelligence and creativity are independent mental abilities. It's possible to be creative although not exceptionally intelligent; most highly intelligent people are not conspicuously creative.

Future research on creativity is likely to dispel another common assertion: that creativity is more likely to be found in the arts than the sciences. Such distinctions are illogical and wrong-headed. A mathematical formula can be as creative as a sonnet. And the application of medical knowledge and skill in the service of arriving at a life-saving diagnosis of a serious but treatable illness is the most marvelous creative expression of all: the rescue of life, from which all creativity emerges, from the implacable clutches of death and oblivion. In my own professional life—divided between writing and the practice of neuropsychiatry—I have experienced in both these very different disciplines the joys attendant on creative expression.

No. Creativity doesn't play favorites, espouses the arts no more nor less than the sciences. Thus it's likely future knowledge about the brain will shed light on the creativity of both the physicist and the dancer, the chemist and the playwright. And why should we expect it to be otherwise, given that the human brain is the progenitor of all creativity?

Arthur Koestler: "Einstein's space is no closer to reality than Van Gogh's sky. The glory of science is not in a truth more absolute than the truth of Bach or Tolstoy, but in the act of creation itself. The scientist's discoveries impose his own order on chaos, as the composer or painter impose his—an order that always refers to limited aspects of reality, and is biased by the observer's frame of reference, which differs from period to period, as a Rembrandt nude differs from a nude by Manet."

CHAPTER THIRTEEN

An Ant in the British Museum

A t this point we are left with a conundrum: If the brain's functioning is based on a modular organization, then what is the basis for the unity of our conscious experience? The fairest answer to that question is that, when it comes to the details, nobody knows. But we have good reason to believe we understand the general principle involved. Most likely unity emerges from the operation of multiple components as the result of neurons interacting at a given level of complexity. This is Roger Sperry's conclusion after a lifetime of studying the neurology of consciousness and personal identity:

> What counts for subjective unity may lie in the way the brain process functions as a unity or entity regardless of the multilevel and multicomponent make-up of the neural event involved. The overall, holistic functional effect could thus determine the conscious experience. If the functional impact of the neural activity has a unitary effect in the upper level conscious dynamics, the subjective experience is unified.

Although that explanation is appealing, at least to me, it does sidestep the question of what level of complexity is sufficient for consciousness. As we have seen, the organization of the human brain does not differ in any fundamentally different way from the brains of the higher mammals (I refer here to cell types, connectivity patterns, and other organizational principles). Indeed, the modular theory of brain function evolved from the study of the macaque and squirrel monkeys. Is the uniqueness of the human brain compared to other animals therefore dependent simply on its size, greater number of cells, and cell connections into networks? If this is accepted, then it seems we must also accept the claim that the difference between ourselves and some of the higher apes is a quantitative rather than qualitative one. Yet despite some initially extravagant claims about the learning of "language" by chimps, claims that have turned out to be greatly exaggerated, no creature other than ourselves is capable of high-level symbolization and abstraction. And there is another reason for discomfort: If consciousness and other "higher" functions are quantitative and dependent on cell connections, why not substitute silicon chips for neurons and construct a cyborg with consciousness and all of our other mental traits?

Much has been written—and a good bit of it nonsense—on the possibility of consciousness evolving from the operation of neural processing and distributed neural networks. Advocates of this approach point out correctly that the brain is so complicated that it cannot be understood outside the context of some guiding theory buttressed by a technology capable of handling enormous amounts of information. Since computers can outperform the human brain in handling mega and giga amounts of information, why not use computers to tell us how the brain works? I would suggest three objections to this theory, the first based on observation gained from living in a real world rather than one comprised of bytes and circuits.

I cannot think of a single example of a simpler level of organization "explaining" the operations in entirety of a more complex one. And it's generally agreed even by the most enthusiastic of computer enthusiasts that the human brain is more complex than any computer created so far. It's facetious, but nonetheless true, to say that the brain created the computer, not the other way round. And like all other human creations, the computer conforms to the specifications and needs of its creator. This isn't to deny that sometimes programmers can devise programs clever enough to outperform the human brain. Computer chess is a good example. Today several programs exist that can beat all but two or three of the best chess players in the world. But the success of these programs tell us very little about how we play chess, because they are not based on refining and improving upon the way humans play chess. In fact, chess computer programs bear little resemblance to human chess-playing ability; they depend on a "brute force" technique of considering more moves and more variations than the brain is capable of handling in the same period of time. But since chess at its highest levels of play is at least partly a matter of intuition, art, inspiration—employ whatever word you wish for a mental process that cannot be completely understood and explained even by the person employing it—the world's best players have had no difficulty so far in beating their silicon-inspired opponents. Based on such examples, which are more the rule than the exception in computer modeling of human cognition, it would therefore require a most extraordinary bootstrap operation for the computer to encompass and explain the workings of its own creator. True, it may help in arriving at certain principles of operation, but it's not likely to provide an complete theory for how it works.

Second, all of the insights we have arrived at so far about parallel processing and modular organization within the brain have come from basic research on the brain and not from computer modeling. Indeed, all of the claims of the neural network experts that the brain can be modeled according to their theory have been made *after* neuroscientists demonstrated parallel processing and multiple interconnected areas of specialization in the brain. Few computer experts admit that neuroscientists had been there first. In their speeches and published writings, the experts on neural processing speak with a sometimes barely concealed contempt for scientists who try to understand the brain by studying the brain. Maybe it's because computer science and mathematics and physics are more difficult subjects to master than neurobiology. If computational scientists believe that—and most of the ones I have spoken to do—then their disdain is at least more understandable, if not necessarily more forgivable. But however computer scientists might wish it otherwise, the brain is organized according to the rules of neurophysiology, not computer science and there is simply no way of getting around that fact. As Semir Zeki puts it, we should be a little careful in accepting theories about the brain ". . . from those whose tools are mathematics and computers rather than brains. The proper way to study the brain is to study the brain."

Finally, the brain is a unique product of evolutionary forces. It evolved in directions that favored successful adaptation to the conditions and forces prevalent on our planet over millions of years. These adaptations, involving life-and-death decisions within incredibly dangerous and threatening environments, could never hope to be simulated in a computer program. If we lived on a different planet, our brain would be organized completely differently and in ways we cannot imagine. It's of interest in this regard to remind ourselves that the human brain has not changed substantially within the past 100,000 years, making it virtually impossible at this point to even identify, much less incorporate into a program, the formative influences on the development of the brain.

Our goal-seeking behavior, along with our capacity to change our goals midstream—these distinguish us from computers, no matter how enticing brain-as-computer analogies may at first appear. A philosopher, J. Beloff, put the matter very well, so let me quote him directly: "A distinguishing attribute of mind . . . is the familiar intentional or purposive aspect of behavior which transforms what otherwise would be a mere sequence of movements into a meaningful action. A machine can go through any sequence of movements which the ingenuity of its inventory will allow but. . . . [t]o equate the cognitive processes of human beings or animals with the information processing of computers is to confuse that which is simulated with its simulation."

I include Beloff's commentary here as a supporting opinion that the brain is not a computer and ergo, we are not biological computers. Joseph Weizenbaum, a professor of computer science at the Massachusetts Institute of Technology, once said to me during a conversation, "We have not so much to fear computers that think, feel, and act as men as we do men who think, feel, and act like computers." Since Weizenbaum said this over a decade ago, I suppose he would be more gender neutral now and would compare "people" rather than "men" to computers. Otherwise the sentence is as valid now as on the day he uttered it. The English philosopher C. D. Broad spoke even more forcefully on the implications of comparing people to machines. "If a man referred to his brother or to his cat as 'an ingenious mechanism' we should know that he was either a fool or a physiologist. No one in practice treats himself or his fellow-man or his pet animals as machines; but scientists seem often to think it their duty to hold in theory what no one outside of a lunatic asylum would accept in practice."

Broad's and Weizenbaum's point (with which I agree entirely, as I expect you can tell) is not just a quibble; if our brain is not like a computer except in a very limited sense, we are not likely to learn a whole lot about how our brain works by confining our efforts to computers. Rather, we have the most to learn by concentrating on what happens within the brain during mental and psychological processes.

All of the special and exciting discoveries about the brain that we have discussed in the previous twelve chapters would not have been possible without the introduction of imaging techniques like the PET scan, capable of displaying in vivid color the brain operations accompanying our thoughts and emotions. Thanks to radioactive emitters tagged to glucose molecules carried in the bloodstream to the brain, measurements can be made of changes in neuronal activity as the subjects carry out various mental tasks. The radioactive tag may be put on a precursor chemical that will go on to form a neurotransmitter; or a drug may be tagged and followed as it courses throughout the brain. In both cases, chemical transformations can be monitored with the facility of an air traffic control person trailing a plane in its flight within an air space. PET holds great promise for learning more about the modular organization of our brain.

Already neuroscientists at Washington University School of Medicine in St. Louis, Missouri, have demonstrated in a brilliant series of experiments the modular nature of language and word processing. For instance, local areas of the visual system of skilled readers passively respond to word and wordlike symbols (i.e., pseudowords like "geel" or "telp") that bear more resemblance to normal English usage than they do to strings of letters that cannot be pronounced (consonant letter strings) or groups of meaningless letterlike symbols arranged as words (false fonts). In essence, the brain knows how to automatically distinguish symbols used for communication from those that have no communication value.

In another experiment by the St. Louis group, subjects were asked to rapidly come up with an appropriate verb when shown a noun (e.g., see "fork," say "eat"). Two findings emerged from this study. First, different circuits are used than for simply repeating the noun displayed on the screen. Second, after as little as fifteen minutes of practice, the PET pattern reverted to the pattern seen with simple repetition (seeing a fork and saying "fork," for example). This corresponded on the behavioral level with the subject's responses: more rapid and more stereotyped with repetition. Marcus Raichle, one of the members of the St. Louis team, believes this study reveals the way the human brain automates and learns complex tasks such as speaking.

"This strategy is employed in normal fluent speech which involves a blend of two strategies, one automatic and the other non-automatic. The two are blended together so that we are aware of the sense of what we are saying rather than every word we say."

Pet scans are also confirming individual learning styles that differ from person to person. And some are more efficient than others. Tests of memory reveal different PET patterns according to whether a person performs well or poorly on tests of memory. The PET scans of bad performers showed activity toward the back in the visual and visual association areas. Raichle speculates that the good performers internally rehearsed the material by silently repeating the material over and over, while the bad performers tried to remember by internally visualizing the words. Such findings may provide the basis for new ways of learning and enhancing mental performance. It's now believed that the brain employs many widespread components which are activated together whenever one reads or remembers. Depending on how these components are coordinated, a variety of different behaviors can result.

"A good analogy is a symphony orchestra with its many sections and different instruments," according to Raichle. "The orchestra does not contain a unique combination of players and instruments that constitute a C-major chord. Rather, it contains many potential combinations that can produce the chord depending on the desired effect. The brain would appear to operate in much the same manner. Our challenge is to understand who the individual players are, what their unique capabilities are and how these capabilities are combined with others to produce the immense flexibility and richness which we appreciate in human behavior."

Already underway are PET studies of highly creative individuals such as artists and professional musicians. As mentioned in Chapter Twelve, this is new ground for neuroscientists who, understandably, have traditionally directed their efforts to the alleviation of the human suffering brought about by brain diseases. Practical considerations have also determined the direction of their efforts: Funding and research grants have almost always been more available for the investigation of illnesses than for creativity. But that emphasis can be expected to change in light of the emerging recognition that the brain is the wellspring of all accomplishments.

Fresh insights into the brain's modular organization are likely to come from technologies, in addition to PET, that involve innovative use of computers combined with new insights from physics to select, manipulate, and process information about the brain. The recent development of the echo-planar MRI (EPI), for instance, greatly enhances the amount of information obtained from standard MRI by using multiple, higher-power, rapidly oscillating magnetic field gradients, more efficient, higher-speed hardware, and better image-processing software. Thanks to its increased imaging speed, EPI provides real-time images of the brain rather than, as with conventional MRI, images obtained over several minutes of measurement. While not quite a "mind-reader," this instrument has the capability of correlating consciousness and awareness with such ongoing events in the brain as changes in the visibility of blood vessels in the brain secondary to the altered behavior of hemoglobin brought about by magnetic fields.

Measurements can now be made of the small magnetic field patterns given off by the neuron's ionic currents. Rather than measuring events in the range of 20 seconds to several minutes, as with PET, magnetoencephalography (MEG) has a temporal resolution in tens of milliseconds, i.e., real time.

Techniques can also be combined, such as EEG and MRI. With the aid of high-power computer programs, the conventional EEG map is superimposed on an MRI image of the subject so that the resulting image provides information about the *location* and *sequence* of extremely rapid occurring events.

But although these new and innovative technologies can provide some insight into the modular functioning of the brain, not everyone is convinced they will shed light on those very important "why" questions that we have discussed over the course of this book. Vernon Mountcastle believes, for instance, that imaging studies can only provide answers to "where" questions at the expense of the "how" and "why" questions.

While I agree in general with Mountcastle about the importance of "why" questions, I wonder if there isn't a paradox here, one of those intellectual mirages that have the power to permanently frustrate our search for self-understanding. In my work with patients, for instance, I've discovered the value of not asking questions like "Why did you do that?" Such inquiries provoke only defensiveness. Much more useful is the simple phrase "What happened *that* such-and-such occurred?" In this way the patient, instead of explaining the essentially unanswerable question "why" she did something that she now regrets, has only to address herself to: "What happened *that* you found yourself in such a position that you felt you had to . . . leave your husband . . . quit your job . . . or whatever other situation arose that brings you for my advice?" I've discovered that if you can shield a person from having to feel directly and personally responsible for something, particularly something he is ashamed or bothered about, he will tell you anything you want to know. And almost always, when I've taken this approach, the "why" eventually emerged indirectly as a secondary benefit. At first I considered this ploy somewhat manipulative; it suggests, at least implicitly, that people exert not an active but a passive influence in shaping their destinies. But when you think about it, there are rarely simple "why" explanations that get us very far in understanding the reasons for our actions, thoughts, emotions, and the contents of our consciousness. "Why"-based explanations of these things are always interpretations rather than causal explanations. This makes it difficult for me to believe that we will ever come up with a satisfying "why" explanation for the brain. If we cannot come up with "why" explanations for our own mental operations as we experience them, then how can we expect to understand "why" the brain functions as it does? Carl Jung must have had something like that in mind when he once compared our search for self-understanding to the efforts of an ant crawling around in the British Museum. Try as it might, that ant will never be capable of understanding the contents of the museum or even "why" the objects are there. I'm not suggesting we drop "why" questions altogether, only that we employ them more sparingly, and with the requisite humility. And there is a good reason for doing this: we really do not have access to all of the operative processes within our own brain.

"The brain doesn't let information become blooming, buzzing confusion. It divvies things up and puts them in bins. And these bins may be different from what you would expect. The brain is not necessarily built the way your mind thinks it is," Barry Gordon, a neurologist and neuropsychologist at Johns Hopkins University, said to me in a conversation about the time I was completing this book. Certainly over the course of this book we have encountered striking examples of this inability on our part to logically or intuitively arrive at the principles of brain organization.

The most striking example of this, and the model for all other forms of inaccessible information processing, is the phenomina of blindsight we mentioned earlier. As you recall, if blind people are asked to "guess" at the location of a light flashed on the wall by an experimenter, they will at first protest that such a determination is impossible but then guess correctly 80 percent of the time. Most fascinating, they will deny seeing anything in their blind visual fields, though if pressed for an explanation, they will speak of a "feeling" about the location of the light.

"Blindsight and other forms of tacit or implied knowledge make up," according to Oxford neuropsychologist Lawrence Weiskrantz, "an epidemic of dissociations: brain processing occurring in the absence of acknowledged awareness." Included here are persons we have encountered during our explorations: amnesic patients who retain some knowledge about events they cannot consciously recall; patients who deny their ability to experience aspects of the sensory world other than vision and yet give correct responses when they are encouraged to "guess"; brain-injured patients who ignore or deny aspects of their world and yet continue to claim nothing is amiss—the dissociation between a capacity and acknowledged awareness takes many forms. Moreover, these dissociated and partial forms of knowledge are of little practical help in the everyday world. All are profoundly disabling: The amnesiac cannot depend on intuitions or feelings; nor can a person afflicted with prosopagnosia avoid the embarrassment that comes from not recognizing his or her children; nor can the patient afflicted with blindsight avoid bumping into obstacles despite the dissociated ability, in forced-choice situations, to guess their location.

But dissociations are not limited to instances of brain damage. Hypnosis and other forms of directed attention produce similar splits in conscious awareness. Elaborate instructions given during the hypnotic trance are later carried out without awareness or memory. Even without hypnosis, dissociations occur during normal mental processing. Think back to something that occurred within the recent past and you can also bring to mind the accompanying emotions. You can't do that with events from your distant past: The feelings are split off, dissociated, from the event (extremely upsetting events, such as witnessing a death or experiencing severe injuries, are an exception and can be recalled for years with painfully vivid intensity, one of the reasons post-traumatic stress disorders are so disabling). Absent "felt" memories of our past, we become as strangers to ourselves: Our photo albums show us the events and people from our past but not the *felt* emotions experienced at the time.

Our brain's modular organization is responsible for these dissociations. And at any given moment we have only to attend to our own inner worlds to experience them. While watching a skilled magician perform or looking at an optical illusion, we experience a conflict between what we know and what we see. We know that the magician cannot actually be sawing the beautiful young woman in half, and yet that's what we see! And when looking at optical illusions on paper, we simultaneously entertain two interpretations of the same configuration of lines. At one moment we see one of them and a moment later the other; if we continue to stare, we may even see the two images come and go so fast they appear present simultaneously.

When making decisions, we inwardly experience the interaction of modules that wax and wane in their power to influence us. At one moment we are certain we have made the correct decision, only to experience a moment later the certainty that the other alternative must be the correct one. Logical paradoxes illustrate the operation—and limitation—of modules governing the operation of reason, logic, and certainty. Logical paradoxes remind us that words and concepts, indeed even the concepts of logic and reason themselves, do not exist in the abstract but are the brain's own creation. Recognition of this fact should lead to a profound humility about our place in the universe along with an enthusiasm, indeed a passion, for learning about our modular brain's principles of operation. And by learning more about the brain we put a new and exciting neurobiological spin on the old Socratic adage, "Know thyself."

GLOSSARY

achromatopsia Disproportionate impairment of the ability to perceive color. The world is typically described as "like a black-and-white movie."

agnosia (from Greek: "without knowledge") Loss of the ability to name or interpret what is seen.

amygdala (from Latin: "almond") A limbic system structure concerned with memory and emotions. The name is based on its shape.

anosognosia The denial of loss of a capacity, such as the loss of the power to move (paralysis) brought on by a stroke.

aphasia Loss of the ability to speak or understand written or spoken language.

apraxia Inability to perform previously learned skilled movements.

autotopagnosia Difficulty comprehending the spoken names of body parts. People afflicted with this retain the ability to point to clothes they are wearing but not to parts of their body.

basal ganglia Several islands of gray matter located beneath the cerebral hemispheres. Included are the caudate, the globus pallidus, the putamen, and others.

brain stem The part of the brain just above the spinal cord and composed of the medulla, the pons and the midbrain. It contains the sensory and motor nerve fibers carrying impulses to and from the brain.

cerebellum The "small brain" located behind the brain stem and below the cerebrum. With the basal ganglia, it coordinates complex movements.

cerebrum The largest part of the brain, consisting of two hemispheres composed of the frontal, parietal, temporal, and occipital lobes; its outer one-eighth-inch layer is called the cerebral cortex, which is the brain area most responsible for thought and other "higher" mental functions.

commissurotomy Surgical transection of the major fiber links (the corpus callosum, the anterior commissure, and the hippocampal commissure) between the two cerebral hemispheres.

corpus callosum   (from Latin: "hard body") A band of nerve fibers uniting the right and left cerebral hemispheres.

hippocampus   A sea-horse–shaped structure, part of the limbic system involved in emotions and memory storage and retrieval.

hypothalamus   Constituting less than 1 percent of the total volume of the brain, this olive-sized structure beneath the thalamus contains a large number of circuits responsible for the regulation of such vital functions as temperature, heart rate, blood pressure, and water and food intake.

limbic system   (from Latin *limbus*: "rim") A group of interrelated structures around the upper brain stem, including the hippocampus, the fornix linking with the hypothalamus, the mammilothalamic tract leading from the mammillary body to the anterior thalamic nuclei, and the cingulum bundle leading to the cingulate gyrus.

pineal   A tiny gland in the center of the brain that plays a role in sleep and sexual maturation.

prosopagnosia   A disproportionate impairment in recognizing faces with preserved ability for person recognition through other channels such as speech.

spinal cord   A cable of nerve cells and their extensions linking the brain to the rest of the body; with the brain forming the central nervous system, to be distinguished from the peripheral nervous system consisting of nerve fibers linking the spinal cord to the muscles and extending from the spinal cord outward to the muscles.

synapse   A narrow, $\frac{1}{1000}$-millimeter space separating the axon terminal of the presynaptic neuron to the dendrite or cell body of a second neuron, the postsynaptic neuron.

thalamus   Located above the hypothalamus, this structure processes all sensory input except smell and forwards it to the cerebral cortex above.

ventricles   Cavities containing cerebrospinal fluid that form an interconnecting system within the brain.

white matter   Bundles of myelin-covered nerve fibers within the central nervous system that provide passage for nerve impulses to and from neurons in the gray matter of the spinal cord and cerebral and cerebellar cortices.

NOTES

CHAPTER ONE: "ON THE ORGAN OF THE SOUL"

Sources for this chapter include: *The Lineaments of Mind* by John Cohen (W. H. Freeman, 1980), especially Chapter 5, "The Mind-Body Relation: Two Paths"; "Mind and Body: Descartes to William James: the Catalogue Accompanying an Exhibition of the National Library of Medicine," 1992; *The Brain Machine* by Marc Jeannerod (Harvard University Press, 1985); *Historical Aspects of the Neurosciences: A Festschrift for Macdonald Critchley*, edit. by Clifford Rose and W. F. Bynum (New York: Raven Press, 1982); *The Oxford Companion to the Mind*, edit. by Richard Gregory (Oxford University Press, 1987).

CHAPTER TWO: DOCTOR MOUNTCASTLE'S CURIOSITY

Background on the visual system taken from "The Visual Image in Mind and Brain" in *Scientific American*, Sept. 92, pp. 6 ff; "Segregation of Form, Color, Movement, and Depth: Anatomy, Physiology, and Perception" by Margaret Livingstone and David Hubel in *Sciences*, May 6, 1988, pp. 740–749; "Art, Illusion and the Visual System" by Margaret Livingstone in *Scientific American*, Vol. 258, Jan. 1988; "Brain mechanisms of visual attention" by Robert H. Wurtz, Michael E. Goldberg, and David Lee Robinson in *Scientific American*, Vol. 247, 1982. "The discovery of the visual cortex" by Mitchell Glickstein in *Scientific American*, Sept. 1988, pp. 118–127. The case of Maureen is taken from *Neurology Chronicle*; she was a patient of Walter Fitzhugh, M.D. The Mountcastle material is from an interview on 21 April 1993 at the Zanvyl Krieger Mind/Brain Institute.

CHAPTER THREE: "A POWER OF WILL"

Background for this chapter from *The Brain Machine* and *The Oxford Companion to the Mind*. The Kenneth Heilman material is from interviews in Antigua, West Indies, 19 and 20 January 1993; also helpful was the piece in *Trends in the Neurosciences*, Jan. 1993, on premotor intention.

NOTES

CHAPTER FOUR: AN EXISTENTIAL ILLNESS

Frontal Lobe Function and Dysfunction, edit. by Harvey S. Levin, Howard M. Eisenberg, and Arthur L. Benton (Oxford University Press, 1991); *Awareness of Deficit After Brain Injury: Clinical and Theoretical Issues*, edit. by George P. Prigatano and Daniel L. Schacter (Oxford University Press, 1991), especially Chapter 2, "Anosognosia Related to Hemiplegia and Hemianopia" by Edoardo Bisiach and Giuliano Geminiani; the Gazzaniga quotes and material are drawn from my review of his book *Nature's Mind* (Basic Books, 1992), which appeared in *The World & I*, Feb. 1993.

CHAPTER FIVE: "IS A BEE SMALLER THAN A HOUSE?"

Material in this chapter based on interviews with Barry Gordon, 30 November 1992; Alfonso Caramazza, 21 April 1993; John Hart, 21 April 1993; "Category-related recognition defects as a clue to the neural substrates of knowledge" by Antonio Damasio, *Trends In Neurosciences*, March 1990 and "Brain and Language" by Antonio Damasio and Hanna Damasio, *Scientific American*, Sept. 1992; also Chapter 6 in *Cognitive Neuropsychology: A Clinical Introduction* by Rosaleen A. McCarthy and Elizabeth K. Warrington (Academic Press, 1990); also *Thought Without Language*, edit. by Lawrence Weiskrantz, especially Chapters 18 and 19. The patients Jennie Powell and Jules Davidoff, Sue Franklin and David Howard are from *Mental Lives: Case Studies in Cognition*, edit. by Ruth Campbell (Oxford: Basil Blackwell, 1992); the Kertesz patient from *Archives of Neurology*, Feb. 1993, p. 195.

CHAPTER SIX: TIME STOPS FOR MR. M

See my essay, "Memories Are Made of This," in *Medical and Health Annual, Encyclopaedia Britannica, 1992*; Chapter Ten, "Preserved Learning Capacity in Amnesia," from *Neuropsychology of Memory*, edit. by Larry R. Squire and Nelson Butters (New York: The Guilford Press, 1984); "Dissociation of Object and Spatial Processing Domains in Primate Prefrontal Cortex" by Fraser A. W. Wilson, Seamus P. O. Scalaidhe, Patricia S. Goldman-Rakic, *Science*, Vol. 260, 25 June 1993.

CHAPTER SEVEN: THE FACE THAT LAUNCHED A THOUSAND SLIPS

The Neuropsychology of Consciousness, edit. by A. D. Milner and M. D. Rugg (New York: Academic Press, 1992) especially Chapter 4, "Face Recognition and Awareness after Brain Injury" by Andrew W. Young and Edward H. F. De Haan. Dr. S. is described in "Developmental memory impairment: faces and patterns" by Christine M. Temple in *Mental Lives: Case Studies in Cognition*, edit. by Ruth Campbell.

CHAPTER EIGHT: BECKY AND LUCY

McCarthy and Warrington, *op. cit.*, Chapter 16, "Problem Solving"; "Disturbance of Self-Awareness after Frontal System Damage" by Donald T. Stuss in *Awareness of Deficit After Brain Injury*, edit. by Prigatano and Schacter; "The Prefrontal Region: Its Early History" by Arthur L. Benton in *Frontal Lobe Function And Dysfunction* by Levin et al.; Chapter 4, "Emotional Disturbances Associated with Focal Lesions of the Limbic Frontal Lobe" in *Neuropsychology of Human Emotion*, edit. by Kenneth M. Heilman and Paul Satz (Guilford Press, 1982).

CHAPTER NINE: THE ALIEN HAND

"The Human Brain: From Dream and Cognition To Fantasy, Will, Conscience, and Freedom" by H. H. Kornhuber in *Information Processing by the Brain: Views and Hypotheses from a Physiological-Cognitive Perspective* (Toronto: Hans Huber Publishers, 1988).

CHAPTER TEN: THE FLIP OF A COIN

Brain and Behavior by Hugh Brown (Oxford University Press, 1976); "The Cognitive Unconscious" by John F. Kihlstrom, *Science*, Vol. 237, 18 Sept. 1987, and "Cognition, Unconscious Processes" in *Neuroscience Year, Supplement 1 to the Encyclopedia of Neuroscience*, edit. by George Adelman Birkhauser (Boston, 1989); "Emotion and the Limbic System Concept" by Joseph E. LeDoux from *Concepts in Neuroscience*, Vol. 2, 1991; "Consciousness, Personal Identity, and the Divided Brain" by Roger Sperry from *Neuropsychologia*, Vol. 22, No. 6, 1984.

CHAPTER ELEVEN: THE PAPEZ CIRCUIT

See *Neuropsychology of Human Emotion* by Heilman and Satz; "Emotion, Neural Substrates" by Orville A. Smith, and "Emotional Short Circuits in the Brain," both in *Neuroscience Year, Supplement 2 to the Encyclopedia of Neuroscience*, edit. by Barry Smith and George Adelman Birkhauser (Boston, 1992); "See No Evil: Blaming the Brain for Criminal Violence," by Richard Restak, *The Sciences*, July/Aug. 1992; *Brain and Behavior* by Hugh Brown; *Mind and Brain: Dialogues in cognitive neuroscience*, edit. by Joseph E. LeDoux and William Hirst (Cambridge University Press, 1986).

CHAPTER TWELVE: THE ILLNESS OF MAURICE RAVEL

"Music, the Brain and Ravel," by Justine Sergent in *Trends in Neurosciences*, May 1993; "The Creative Brain" by Richard Restak in *Creativity*, edit. by John Brockman Touchtone (New York: Simon and Schuster, 1993).

NOTES

CHAPTER THIRTEEN: AN ANT IN THE BRITISH MUSEUM

The Raichle materials are from his lectures at the American Academy of Neurology, New York City, 1993.

INDEX

INDEX